Benvenuto (Ed.)
Green Chemical Processes
Green Chemical Processing

Green Chemical Processing

Edited by
Mark Anthony Benvenuto

Volume 2

Green Chemical Processes

Developments in Research and Education

Edited by
Mark Anthony Benvenuto

DE GRUYTER

Editor
Prof. Dr. Mark Anthony Benvenuto
Department of Chemistry and Biochemistry
University of Detroit Mercy
4001 W. McNichols Road
Detroit, MI 48221-3038
USA

ISBN 978-3-11-065251-2
e-ISBN (PDF) 978-3-11-044592-3
e-ISBN (EPUB) 978-3-11-043723-2
Set-ISBN 978-3-11-044593-0
ISSN 2366-2115

The articles in this book have been previously published in the journal *Physical Sciences Reviews* (ISSN 2365–659X).

Library of Congress Cataloging-in-Publication Data
A CIP catalog record for this book has been applied for at the Library of Congress.

Bibliographic information published by the Deutsche Nationalbibliothek
The Deutsche Nationalbibliothek lists this publication in the Deutsche Nationalbibliografie; detailed bibliographic data are available on the Internet at http://dnb.dnb.de.

This volume is text- and page-identical with the hardback published in 2017.
Typesetting: Integra Software Services Pvt. Ltd.
Printing and binding: CPI books GmbH, Leck
Cover image: scyther5/iStock/Thinkstock
♾ Printed on acid-free paper
Printed in Germany
www.degruyter.com

Contents

List of contributing authors

Chapter 2
George Bodner
Purdue University
Department of Chemistry
560 Oval Drive
West Lafayette, IN 47907-2084
USA
gmbodner@purdue.edu

Chapter 3
Larry Kolopajlo
Eastern Michigan University
Department of Chemistry
501T Mark Jefferson Science Complex
Ypsilanti, MI 48197
USA
lkolopajl@emich.edu

Chapter 4
Jonathan Stevens
University of Detroit Mercy
4001 W. McNichols Rd.
Detroit, MI 48221-3038
USA
stevenje@udmercy.edu

Chapter 5
Serenity Desmond
University of Illinois at Urbana-Champaign
Department of Chemistry
601 S. Mathews Avenue
Urbana, IL 61801
USA
sdesmond@illinois.edu

José Andino Martinez
University of Illinois
Department of Chemistry
600 S. Mathews Avenue
Urbana, IL, 61801
USA
andinoma@illinois.edu

Christian Ray
University of Illinois
Department of Chemistry
600 S. Mathews Avenue
Urbana, IL, 61801
USA
crray@illinois.edu

Chapter 6
Daniel Pharr
Virginia Military Institute
Lexington, VA 24450
USA
pharrdy@vmi.edu

Chapter 7
David Consiglio
Southfield High School for the Arts and Technology
24675 Lahser Road
Southfield, MI
USA
davecons@gmail.com

Chapter 8
Steven Kosmas
Grosse Pointe North High School
707 Vernier Road
Grosse Pointe Woods, MI 48236
USA
Steven.Kosmas@gpschools.org

Mark Benvenuto

1 Incorporating green chemistry into education

Abstract: This volume includes several different author voices discussing how green chemistry is utilized in our educational system. American Chemical Society Governing Board member George Bodner discusses this at the college level, while several other authors present how they incorporate one or more of the principles of green chemistry into college and high school chemistry classes [1].

Keywords: green chemistry, college and high school chemistry classes, education

1.1 Introduction

> "Education is the most powerful weapon which you can use to change the world."
> Nelson Mandela

Green chemistry is undoubtedly about changing the world, changing it for the better, to make it a sustainable home for humanity now and in the future. Yet there is still a great deal to do if we wish to construct a future in which chemicals and materials produced on an industrial scale are produced in alignment with the principles of green chemistry, especially in terms of creating minimal amounts of waste or by-products. A thorough grounding of green chemical principles into our educational structure will indeed be a powerful weapon in this continuing struggle to not only change but also improve our world.

While what is now called green chemistry can trace itself back to the Brundtland Report [2], and early efforts on the part of the US Environmental Protection Agency (EPA) and the American Chemical Society [3, 4]. There have certainly been numerous other efforts, sometimes independently undertaken, sometimes more localized, in which national or multi-national organizations [5–11] as well as academic institutions and individuals have tried to bring the 12 principles of green chemistry to the classroom [12–17]. The argument is very easy to make that if we wish to turn the chemical enterprise in all its forms to a greener posture, that although efforts can begin anywhere, education is one obvious, logical starting point.

In our first volume, Professor Sarah Green made the interesting comment that when she first heard of green chemistry, as a graduate student, she felt that the idea was admirable, but that the entirety of the challenge would be met by the time she was through with her education – and was disappointed to find that it was far from so when she had obtained her degree. There is a great deal of both wisdom and insight in such a comment. It appears obvious to those who have embraced the principles of

https://doi.org/10.1515/9783110445923-001

green chemistry that this should be a better means of doing business. Yet changing the minds, the behavior, and the economically-driven actions of a huge number of people – such as those in the enormous enterprise we call the chemical industry – is itself an enormous task. Industry will not look for green changes if there is no incentive, and economic incentives are the most enticing. Government will not request, require, or pursue changes if the people who constitute the governmental apparatus are unaware that better alternatives may exist. And new employees in any aspect of our science – educational, corporate, or government – will not think to pursue changes to a greener posture if they are unaware that such are even conceivable. Hence, the education of our young becomes a matter of planting the seeds for a new generation and a new way of thinking about what we produce and how we produce it.

Education means different things to different people. For many, it is the formal process that stops either at the end of high school or the end of a four-year college education. Indeed, there are several education-based organization with developed, quite detailed green chemistry components to them (with apologies to those not included in our references) [11–17]. But to corporate managers, for example, education may very well mean the continuing improvement of their work force. This form of life-long education for chemists and chemical engineers who are already employed can definitely include ideas, projects, and components relating to green chemistry. Curiously though, there do not appear to be organizations devoted to linking green chemistry to this level of continuing, life-long education. Rather, material from those organizations devoted to more traditional education is generally incorporated here.

1.2 This volume

Welcome to the second volume of this evolving series on green and sustainable chemistry, chemical practices, and chemical processes.

Perhaps obviously, the current volume focuses on educational efforts in green chemistry and spans a variety of authors, from Distinguished Professor George Bodner, as well as several other professors and teachers, to two "foot soldiers" among a legion of educators who are high school teachers, and thus who are teaching green chemistry at what one can argue is the earliest point in a student's education.

1.2.1 The quadruple bottom line

We are indeed honored to have Professor George M. Bodner, the Arthur E. Kelly Distinguished Professor of Chemistry, Education and Engineering at Purdue University, and a member of the American Chemical Society Board of Directors, present a chapter with this intriguing title. He gives valuable insights into the "economic, social, and environmental domains" of green chemistry education and couples this with bringing relevance of the subject to the undergraduate curriculum.

1.2.2 Green chemistry in the Middle East

Professor Larry Kolopajlo has submitted a fascinating chapter on green chemistry education and its continuing evolution in the Middle East. The area of the world that is so completely dominated by petro-chemistry is not immediately thought of as being a region from which novel ideas in green chemistry originate. Yet Dr. Kolopajlo does an excellent job of presenting the often subtle, low-key, but very important efforts being made by educators in this area.

1.2.3 Virtually green chemistry

It is Professor Jon Stevens who has given us a thought-provoking chapter about the role computational chemistry plays in the field of green chemistry and in chemical education efforts. He presents several reasoned arguments as to why chemistry done via computer represents a logical starting point for any efforts at chemical synthesis, and how such efforts can be incorporated into the curriculum.

1.2.4 Greening the teaching lab

A team from the University of Illinois has presented several very practical examples of how a greening of the curriculum can benefit both the students and the faculty at an educational institution. The team focuses on making the general chemistry labs greener in terms of making them safer, making them a better educational experience, and at the same time making them more cost-effective. The lattermost is a benefit that universities and colleges of all sizes can embrace.

1.2.5 Surfactants versus solvents

Professor Dan Pharr has presented us with a chapter that provides a detailed look at the use of surfactants in a wide variety of chemical roles and examines how their "greenness" can be measured against more traditional solvents. This is an area of chemistry that has seen some study, but that still has numerous possibilities for advances in the future. Additionally, the study of surfactants is an area within the undergraduate curriculum that has traditionally been slighted in favor of other subjects deemed more important. This is ironic, since a large segment of the chemical industry that will ultimately employ chemists and chemical engineers with the bachelor's degree uses surfactants in many ways, and thus requires graduates who have an understanding of such materials. A knowledge of how these can be used to make environmental and economic improvements in chemical processes will surely be invaluable in the future.

1.2.6 Bio-sources and energy production

We are pleased and honored to have two high school teachers write in this volume. The first, Mr. David Consiglio, speaks on bio-fuels and bio-energy. Importantly, he emphasizes the numerous ways in which bio-fuels and energy can be discussed and utilized as topics in the high school chemistry classroom.

1.2.7 The green chemistry growth mindset

Another high school teacher who gives us insights in this volume is Mr. Steven Kosmas, who discusses not just the green chemistry topics and examples he uses in his high school classroom, but also examines what he refers to as a green chemistry growth mindset. Mr. Kosmas emphasizes that the development of this mindset, this way of thinking about all chemical processes, is an educational tool that can certainly go with students long after they have graduated from a formal educational setting.

In their chapters, both high school teachers present examples they have been able to incorporate into their classes, and thus present information that may be extremely valuable in other academic settings. High school is the final phase of a person's education that is required by the governments of most nations. It may therefore be the only formal setting that many citizens have in which the broad ideas of green chemistry are presented to them. For this reason, what high school teachers actually teach is extremely important. We consider these two chapters to be valuable additions to this volume.

1.3 Summary

Green chemistry in education is perhaps obviously important in training the next generation, whether they are to be scientists and engineers, or simply well-informed citizens who are knowledgeable about how their country and world works. High school and college students can easily be exposed to the 12 principles when under the tutelage of caring, intelligent, well-informed faculty members.

We have been fortunate to be able to include here several chapters that approach green chemistry in education from very different points of view. Being able assemble these chapters in one book is quite important we believe, since it is today's students who will be tomorrow's chemists and other scientific professionals.

References

[1] Twelve principles of Green Chemistry. Website (Accessed 17 June 2016). Available from: https://www.acs.org/content/acs/en/greenchemistry/what-is-green-chemistry/principles/12-principles-of-green-chemistry.html

[2] Our common future: the world commission on environment and development. Oxford University Press, 1987, 019282080X.

[3] US Environmental Protection Agency. Green Chemistry. Website (Accessed 15 January 2017). Available from: https://www.epa.gov/greenchemistry

[4] ACS Green Chemistry Institute. Website (Accessed 15 January 2017). Available from: https://www.acs.org/content/acs/en/greenchemistry.html

[5] Green Chemistry Section of the South African Chemical Institute. Website (Accessed 15 January 2017). Available from: http://www.saci.co.za/GreenChemistry/default.html

[6] OECD. Website (Accessed 15 January 2017). Available from: http://www.oecd.org/env_sustainablechemistry_platform/

[7] Green Centre Canada. Website (Accessed 15 January 2017). Available from: https://www.greencentrecanada.com/

[8] SusChem, European Technology Platform For Sustainable Chemistry. Website (Accessed 15 January 2017). Available from: http://www.suschem.org/

[9] Canadian Green Chemistry Network. Website (Accessed 15 January 2017). Available from: http://www.greenchemistry.ca/

[10] Chemistry Education Association. Website (Accessed 15 January 2017). Available from: http://www.cea.asn.au/

[11] Michigan Green Chemistry Clearinghouse. Website (Accessed 15 January 2017). Available from: https://www.migreenchemistry.org/toolbox/directory/

[12] Green Chemistry at the University of Oregon. Website (Accessed 15 January 2017). Available from: http://greenchem.uoregon.edu/

[13] Hjeresen DL, Schutt DL, Boese JM. Green Chemistry and Education. J Chem Ed 2000;77:1543–1547. [Internet]. Accessed 15 January 2017. Available from: http://www.d.umn.edu/~rdavis/courses/che3791/Green/Papers/Education.pdf

[14] University of Toledo, School of Green Chemistry and Engineering. Website (Accessed 15 January 2017). Available from: http://www.utoledo.edu/nsm/sgce/.

[15] University of York, Green Chemistry Centre of Excellence. Website (Accessed 15 January 2017). Available from: http://www.york.ac.uk/chemistry/research/green/

[16] Carnegie Mellon, Institute for Green Science. Website (Accessed 15 January 2017). Available from: http://igs.chem.cmu.edu/

[17] CGEdNet. Green Chemistry Education Network. Website (Accessed 15 January 2017). Available from: http://cmetim.ning.com/

George M. Bodner

2 The quadruple bottom line: the advantages of incorporating Green Chemistry into the undergraduate chemistry major

Abstract: When the author first became involved with the Green Chemistry movement, he noted that his colleagues in industry who were involved in one of the ACS Green Chemistry Institute® industrial roundtables emphasized the take-home message they described as the "triple bottom line." They noted that introducing Green Chemistry in industrial settings had economic, social, and environmental benefits. As someone who first went to school at age 5, and has been "going to school" most days for 65 years, it was easy for the author to see why introducing Green Chemistry into academics had similar beneficial effects within the context of economic, social and environmental domains at the college/university level. He was prepared to understand why faculty who had taught traditional courses often saw the advantage of incorporating Green Chemistry into the courses they teach. What was not as obvious is why students who were encountering chemistry for the first time were often equally passionate about the Green Chemistry movement. Recent attention has been paid, however, to a model that brings clarity to the hitherto vague term of "relevance" that might explain why integrating Green Chemistry into the undergraduate chemistry classroom can achieve a "quadruple bottom-line" for students because of potentially positive effects of adding a domain of "relevance" to the existing economic, social, and environmental domains.

Keywords: Green Chemistry, education, sustainability, Sustainable Development

2.1 Introduction

This chapter will describe the effect of the Green Chemistry movement on those of us who devote a considerable amount of our professional careers to trying to improve the way we "teach chemistry," as opposed to recommendations in this volume about improving the way people "do chemistry." In doing so, it will trace the genesis of the Green Chemistry movement; differentiate between the concepts of Green Chemistry and Sustainability (or Sustainable Development); examine the philosophy upon which the incorporation of Green Chemistry into the chemistry curriculum might be based; and try to justify the author's belief that Green Chemistry not only can, but should, be incorporated into every chemistry course from students' first exposure in the K-12 classroom to the last graduate level course they take on the way to a Ph.D. in some aspect of what can be referred to as the "chemical enterprise."

https://doi.org/10.1515/9783110445923-002

2.2 The Green Chemistry movement as a community of practice

When a member of the ACS Board of Directors, the author was often been involved in discussions of why individuals become members of the Society. Access to "information" within the context of our journals, SciFinder and other products of the Chemical Abstract Service, and technical meetings/conferences were always high on the list. But it has been interesting to note that "opportunities to network" is toward the topped the list, as well.

A useful model for understanding the idea of networks of professionals working in a given sub-discipline or domain of the chemical enterprise might be the concept of "Community of Practice" [1], which can be defined as "... groups of people who share a concern, a set of problems, or a passion about a practice and who deepen their knowledge and expertise by interacting on an ongoing basis" [2]. These groups often evolve through interactions that occur during professional discourse and collaboration and have the effect of making members of the community more effective as they share knowledge [3]. As anyone who works on research projects that involve either undergraduate or graduate students soon learns, experts in a discipline become more productive by helping/mentoring others [2]. Within the context of teaching/learning, students' grasp of both the content of their discipline and the process of reasoning about this content improve as they learn to communicate ideas and engage each other in dialogue.

Research has shown the positive effects of collaborative inquiry on the acquisition of scientific ways of knowing and reasoning among language-minority students in middle and high school [4]. This work has also noted that the students' reasoning skills improve as they learn to communicate ideas and engage each other in dialogue [4]. As Brown, Collins and Duguid [5] concluded, this seems to occur best within the context of authentic problems or tasks; tasks that arise in the course of a shared need, concern or problem. In theory, getting students involved in implementing Green Chemistry can provide the context of an authentic problem or task.

2.3 Differentiating between Green Chemistry and Sustainable Development

As someone who has been an associate editor of four journals that publish what has become known as discipline-based educational research (DBER), the author is unusually sensitive to the importance of the phrase: "For the purpose of this study, we will assume that ...". The idea of explicitly stating assumptions upon which a study will be based is by no means unique to DBER; many fields share a problem that results from having different authors use a particular word or phrase in different ways, with different fundamental assumptions upon which the use of this word or phrase is based. So let me start by noting that some authors seem to conflate the terms *Green Chemistry* and *Sustainable Development* or *Sustainability*. For the purpose of this chapter, the author will assume that these terms should not be used interchangeably. For the sake of argument, the author will use the definition of Green Chemistry proposed by the U.S. Environmental Protection Agency: "... the design of chemical products and processes that reduce or eliminate the generation of hazardous substances" [6].

Having argued that we need to distinguish between Green Chemistry and Sustainable Development, we now need a working definition of the latter term. The International Institute for Sustainable Development [7] has noted that the most frequently quoted definition of this term appeared in the report known as *Our Common Future* from the World Commission on Environment and Development [8]:

> Sustainable development is development that meets the needs of the present without compromising the ability of future generations to meet their own needs. It contains within it two key concepts:

- The concept of needs, in particular the essential needs of the world's poor, to which overriding priority should be given.
- The idea of limitations imposed by the state of technology and social organization on the environment's ability to meet present and future needs.

We can now understand why the Royal Society of Chemistry (RSC) website suggests that green chemistry "can be considered as chemists aspiring to the principles of Sustainable Development" [9]. And why others have called for "tools to teach and learn about green chemistry, in order to create a sustainable future" [10].

2.4 Genesis of the Green Chemistry movement

As the author has noted elsewhere [11], the Green Chemistry movement can be traced back to the Pollution Prevention Act of 1990, which represented a fundamental change in national policy. For the first time, the EPA was mandated to focus on preventing pollution at its source rather than treating pollutants once they were formed. Section 6602(b) of the Pollution Prevention Act set forth the following hierarchical policy for the United States [12]:

- Pollution should be prevented or reduced at the source whenever feasible.
- Pollution that cannot be prevented should be recycled in an environmentally safe manner whenever feasible.
- Pollution that cannot be prevented or recycled should be treated in an environmentally safe manner whenever feasible.

- Disposal or other release into the environment should be employed only as a last resort and should be conducted in an environmentally safe manner.

When the Pollution Prevention Act of 1990 was signed into law, Paul Anastas was the head of the Industrial Chemistry Branch at the EPA. In 1991, Anastas invented the term "Green Chemistry" to describe a different way of thinking about how chemistry and chemical engineering are done. Although this is not often recognized, the philosophy behind the Green Chemistry movement was not limited to recognizing potential hazards. As it has been explained on the American Chemical Society (ACS) website [13]:

> It's important to note that the scope of these green chemistry and engineering principles go beyond concerns over hazards from chemical toxicity and include energy conservation, waste reduction, and life cycle considerations such as the use of more sustainable or renewable feedstocks and designing for end of life or the final disposition of the product.

In an interview published in *Nature* in 2011, Anastas was asked: "How did you come up with the name 'Green Chemistry'?" He responded: "... they think I'm joking when I say, well, green is the colour of nature, but in the United States green is also the colour of our money. It's always been about how you meet your environmental and economic goals simultaneously" [14].

A major step forward in the Green Chemistry movement occurred in 1995, when President Clinton created the Presidential Green Chemistry Challenge Awards. These awards did more than give credit where it was due, they served as models for others to follow and a way of tracking progress. For a technology to be considered Green Chemistry, it must be more environmentally benign, more economically viable, and functionally equivalent to or outperform existing alternatives [15]. According to the Warner-Babcock Institute for Green Chemistry website, only 10 % of current technologies are "environmentally benign," while another 25 % "could be made benign relatively easily" [15]. This means there is a great deal of room (65 %) for further innovation.

2.4.1 12 principles of Green Chemistry

John Warner is recognized as a co-founder of the Green Chemistry movement. Warner worked with Anastas to create the Presidential Green Chemistry Challenge Awards and served as co-author with Anastas of the seminal textbook in the field: *Green Chemistry: Theory and Practice* [16], in which the authors summarized a set of 12 guiding principles. The original wording of these guiding principles can be found on various websites, but the language with which they are described has been made a little less technical in recent years. As described on the EPA website [17] that defines the basics of green chemistry, for example, they are:

1. **Prevent waste**: Design chemical syntheses to prevent waste. Leave no waste to treat or clean up.
2. **Maximize atom economy**: Design syntheses so that the final product contains the maximum proportion of the starting materials. Waste few or no atoms.
3. **Design less hazardous chemical syntheses**: Design syntheses to use and generate substances with little or no toxicity to either humans or the environment.
4. **Design safer chemicals and products**: Design chemical products that are fully effective yet have little or no toxicity.
5. **Use safer solvents and reaction conditions**: Avoid using solvents, separation agents, or other auxiliary chemicals. If you must use these chemicals, use safer ones.
6. **Increase energy efficiency**: Run chemical reactions at room temperature and pressure whenever possible.
7. **Use renewable feedstocks**: Use starting materials (also known as feedstocks) that are renewable rather than depletable. The source of renewable feedstocks is often agricultural products or the wastes of other processes; the source of depletable feedstocks is often fossil fuels (petroleum, natural gas, or coal) or mining operations.
8. **Avoid chemical derivatives**: Avoid using blocking or protecting groups or any temporary modifications if possible. Derivatives use additional reagents and generate waste.
9. **Use catalysts, not stoichiometric reagents**: Minimize waste by using catalytic reactions. Catalysts are effective in small amounts and can carry out a single reaction many times. They are preferable to stoichiometric reagents, which are used in excess and carry out a reaction only once.
10. **Design chemicals and products to degrade after use**: Design chemical products to break down to innocuous substances after use so that they do not accumulate in the environment.
11. **Analyze in real time to prevent pollution**: Include in-process, real-time monitoring and control during syntheses to minimize or eliminate the formation of byproducts.
12. **Minimize the potential for accidents**: Design chemicals and their physical forms (solid, liquid, or gas) to minimize the potential for chemical accidents including explosions, fires, and releases to the environment.

These principles can be divided into two broad categories: (1) reducing risk and (2) minimizing the environmental footprint. If asked to do so, the author could discuss in great detail the implications of many of these guiding principles in terms of providing new ways to think about the process of both “doing chemistry” and “teaching chemistry.” But for now, he would like to emphasize the connection between the social and/or environmental benefits of Green Chemistry as outlined in the 12 principles listed above and the idea of explicitly thinking about our work as chemists within the general context of both "safe practice" and "prudent practice." He would also like to propose an intrinsic connection between the general principles of Green Chemistry and ethical practice that results, in part, because of the intrinsic connection between the basic tenants of chemical health and safety and ethical practice in the chemistry classroom/laboratory.

2.4.2 12 principles of Green Engineering

Anastas and Zimmerman [18] went on to define green engineering as "the development and commercialization of industrial processes that are economically feasible and reduce the risk to human health and the environment." The guiding principles for green chemical engineering were defined as follows:

1. **Inherent Rather Than Circumstantial**: Designers need to strive to ensure that all materials and energy inputs and outputs are as inherently nonhazardous as possible.
2. **Prevention Instead of Treatment**: It is better to prevent waste than to treat or clean up waste after it is formed.
3. **Design for Separation**: Separation and purification operations should be designed to minimize energy consumption and materials use.
4. **Maximize Efficiency**: Products, processes, and systems should be designed to maximize mass, energy, space, and time efficiency.
5. **Output-Pulled Versus Input-Pushed**: Products, processes, and systems should be "output pulled" rather than "input pushed" through the use of energy and materials.
6. **Conserve Complexity**: Embedded entropy and complexity must be viewed as an investment when making design choices on recycle, reuse, or beneficial disposition.
7. **Durability Rather Than Immortality**: Targeted durability, not immortality, should be a design goal.
8. **Meet Need, Minimize Excess**: Design for unnecessary capacity or capability (e. g., "one size fits all") solutions should be considered a design flaw.
9. **Minimize Material Diversity**: Material diversity in multicomponent products should be minimized to promote disassembly and value retention.
10. **Integrate Material and Energy Flows**: Design of products, processes, and systems must include integration and interconnectivity with available energy and materials flows.
11. **Design for Commercial "Afterlife"**: Products, processes, and systems should be designed for performance in a commercial "afterlife."
12. **Renewable Rather Than Depleting**: Material and energy inputs should be renewable rather than depleting.

2.4.3 Implications of the use of the term "design" in the guiding principles

In 2005, the author worked with colleagues from the College of Engineering at Purdue to create what has become the School of Engineering that was built, in part, upon the framework of Chemical Education graduate program at Purdue that had been created roughly 25 years earlier. When he first started working with colleagues in engineering he noticed that they repeatedly tried to differentiate themselves from scientists by claiming: "We do 'design'." His reaction to this argument was to probe their misunderstanding of the nature of science, arguing that "design" was a critical component of the practice of "doing science." Consider the compound whose structure is shown in Figure 2.1, which was the product of a total synthesis proposed by an organic chemistry graduate student during an oral exam in which the author took part.

If the creation of the multistep process by which one would synthesize this substance is not an exercise in "design," then the author has no understanding of what that word might mean. The other example he used as an illustration of the concept of design in the sciences was to ask his colleagues in engineering

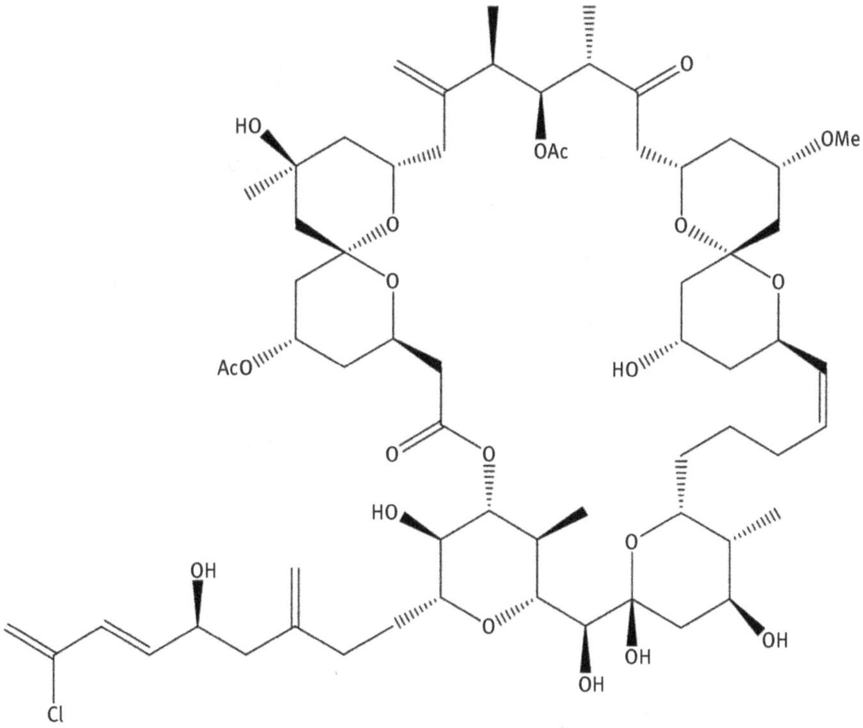

Figure 2.1: Spongistatin 1: a macrocyclic lactone with broad-spectrum antifungal activity.

to consider the Large Hadron Collider at CERN that was involved in the successful search for the Higgs boson, and asked: Isn't the process whereby literally thousands of physicists were involved in the construction of this apparatus an exercise in "design?" The author's research group went so far as to carry out a study comparing the meaning of the term "design" in such diverse fields as choreography, creative writing, organic chemistry, and engineering [19]. This temporary digression from the flow of discourse in this chapter was inserted as an excuse to remind the reader of the frequency with which the term "design" can be found in the principles of both Green Chemistry and Green Engineering.

2.5 The ACS Green Chemistry Institute (ACS GCI®)

The Green Chemistry Institute® was created as a not-for-profit corporation in 1997 that became part of the American Chemical Society four years later. The goal of the ACS CGI® is to be "the premier agent of change providing the knowledge, expertise and capabilities to catalyze the movement of the chemical enterprise toward sustainability through the application of green chemistry principles" [20]. Under the heading, "What is Green Chemistry" [21] the ACS website makes a series of important points: "Green chemistry is not politics; it is not a public relation ploy; nor is it a pipe dream. It is a field described as open for innovation, new ideas, and revolutionary progress. Most importantly, it is the future of chemistry."

During one of the annual Green Chemistry and Engineering Conferences, the author was involved in a conversation among a group of industrial chemists who proposed the notion that Green Chemistry is a global effort to solve what are inherently local problems. If this is true, they argued, different geographical regions will have to take different approaches to solving what might appear to be similar problems. Thus, they argued, green chemistry jobs may be unique because they cannot be outsourced. As recently noted in *Chemical and Engineering News* (C&EN), "The 2011 Pike Research report estimated that the market for chemicals produced through green chemistry approaches will reach nearly $100 billion by 2020, and many industry sectors are vying for their share" [22]

The very existence of this volume is evidence of the commitment of many aspects of what is called "the chemical enterprise" to Green Chemistry. One of the examples the author uses in his own courses is based on the well-known case study involving the synthesis of ibuprofen [23]. He points out the obvious to his students: If you need to make 10 grams of this substance, the traditional idea of "percent yield" is all one needs, and, although one would like to get a better yield, the traditional six-step synthesis of ibuprofen that is characterized by a 40 % yield is "workable." But, once we understand that 30 million pounds of this substance need to be synthesized each year, the production of 45 million pounds of waste is obviously unacceptable. Students therefore readily appreciate the advantage of the three-step synthesis with an atom economy of as much as 99 percent that was developed by the BHC Company, a joint venture of Hoechst Celanese and the Boots Company. They recognize that this process generates more of the desired product, in less time, with less energy, and therefore less impact on the economy. Without having fully integrated the guiding principles of Green Chemistry into their thought process, they can appreciate why the pharmaceutical industry is a particularly good example of the implementation of green chemistry.

Another example from industry that is easily incorporated into introductory classes is the story of sildenafil citrate. Students readily accept the thought that the first priority of a company might be getting a product "out the door." And they readily understand that, once this has happened, the next priority might be improving the efficiency. As noted by Poliakoff [24], the original synthesis of the active ingredient in the "blue pill" sold by Pfizer once consumed 1350 liters of solvent per liter of product. Today, it consumes only 6 liters of solvent per kilogram of product. So they are willing to accept the concept of the "triple bottom line;" the idea that Green Chemistry can have strong economic benefits as well as both social and environmental benefits.

Having set the basis for thinking about Green Chemistry in industry and the reasons why many companies are willing to pay the cost of joining the Pharmaceutical or Industrial Roundtables organized by the ACS GCI™, the author has been able to bring into the discussion more subtle examples, such as the use of 2-methyltetrahydrofuran. He starts by noting that companies often have a "Green Chemistry" component on their website. He then points out that the first example of Green Chemistry on the Aldrich Chemical website [25] was a substance he first learned about by talking to industrial chemists at one of the Green Chemistry and Engineering conference: 2-methyltetrahydrofuran (Figure 2.2).

Having distinguished between the structures of this compound and THF, he asks the question: What makes this solvent "green"? It is obtained from renewable resources such as corncobs or the fibrous material that remains after sugarcane or sorghum stalks are crushed. It is also safer than THF because it is less likely to form peroxides. It is both a polar solvent and an aprotic solvent, so Grignard reagents are

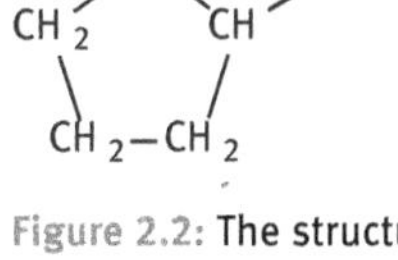

```
      O        CH 3
   /     \    /
CH 2      CH
  \       /
  CH 2—CH 2
```

Figure 2.2: The structure of 2-methyltetrahydrofuran.

more soluble in 2-MeTHF than THF. Because it has only a limited solubility in water, 2-MeTHF is much easier to separate and recover, thereby reducing the waste stream. Finally, it has a low heat of vaporization, which means that less energy is consumed during distillation and recovery.

2.6 Green Chemistry in the classroom

Regardless of whether one is an industrial chemist worrying about how to make several million pounds of a substance or an instructor thinking about new ways of overcoming the difficulties associated with the teaching and learning of chemistry, Green Chemistry does not change the facts or principles of chemistry. It just asks us to bring a new dimension into our thought process; a dimension in which:

- Chemistry is actively integrated into the world around us.
- A priority is placed on minimizing potentially negative effects on the environment.
- Where both practicing chemists and students taking chemistry for the first (and perhaps last) time are actively involved in thinking about safe practice,
- Where minimizing waste and maximizing atom-economy become more important concepts than the traditional driving factor of "percent yield."
- Where the notion of limited resources is understood and resources that are used are renewable.

The Green Chemistry movement reminds us of something we ask students to do that is often not done by practicing chemists: Writing a balanced equation so that attention is focused on not just the starting materials and target of the reaction, but the waste generated in the course of the reaction.

The author believes that examples of green chemistry should be incorporated into every chemistry course, at any level; from the introduction to chemistry taught in the K-12 classroom through the last course graduate students take to earn a Ph.D. in chemistry or chemical engineering. A clear understanding of the philosophy upon which the Green Chemistry movement is based is essential for the handful of students from the general population who will go on to pursue careers in the chemical enterprise. But it is also useful for the vast majority of students who will pursue careers in other fields because it can provide a basis for connecting the concepts and facts we ask them to learn to concrete examples that are likely to be more "relevant" than the traditional exercises that have dominated our courses for so many years.

2.6.1 Beyond benign

At about the same time that proponents were first formalizing what became the Green Chemistry movement, a book was published called *Way Past Cool* [26]. The idea upon which it was based is that it is easy to be "cool," but much more difficult to be "way past cool." Within this context, let's turn to the work of an organization known as "Beyond Benign." (Not just "benign," but "beyond benign.") As noted on their website [27], the mission of Beyond Benign is to provide "... scientists, educators and citizens with the tools to teach and learn about green chemistry, in order to create a sustainable future," and its vision is "to revolutionize the way chemistry is taught to better prepare students to engage with their world while connecting chemistry, human health and the environment" [10]. Beyond Benign was created by John Warner to provide an approach for scientists involved in Green Chemistry to reach out to the public because of his belief that:

> Green Chemistry provides the perfect platform for communicating the importance of science in providing solutions for many of society's challenges because Green Chemistry inherently minimizes the impacts of science on the environment and it is a sustainable approach to chemistry. The relationship between Green Chemistry and the environment provides a uniquely positive, solutions-based starting point for encouraging younger students, who are greatly interested in the environment, to consider positive contributions they can make in any scientific field [26].

Historically, Beyond Benign focused on three aspects of the Green Chemistry movement: (1) the K-12 curriculum, (2) community outreach, and (3) workforce development. The K-12 efforts are based on the idea that the concepts of Green Chemistry and Sustainability will be essential knowledge for all future scientists and educated citizens. Beyond Benign therefore produces lesson plans, curriculum materials, and training programs for the K-12 community.

The website for Beyond Benign has sections devoted to colleges and universities, a Fellows program, and both online courses for educators and online workshops. It provides links to Green Chemistry resources on the web; publications; consumer guides; a "how to" guide to incorporating Green Chemistry concepts and activities into a year-long chemistry course; lessons "for teachers, by teachers" for use in middle-school or high-school classrooms; information about a Green Math, Engineering and Technology project; introductions to biotechnology for use by either middle-school or high-school teachers; access to a peer reviewed journal *Green Chemistry Letters & Reviews*; materials "by professors, for professors" targeted toward community colleges; and information about workforce development proposing the use of Green Chemistry as a tool for regional economic development and job creation.

In different contexts, and using slightly different language, the co-founders of the Green Chemistry movement – Paul Anastas and John Warner – have each proposed a metric upon which to judge the success of the Green Chemistry movement: It will be successful, they argue, when the term Green Chemistry disappears because it will have become integrated into the practice of either "doing chemistry" or "teaching chemistry.

No-one would argue, however, that we are close to having achieved that goal. As a step in the right direction, a program known as the *Green Chemistry Commitment*

[28] has been created to bring together a Community of Practice that would be built around the shared goals of: (1) expanding the number of practitioners of green chemistry, (2) increasing departmental/institutional resources devoted to green chemistry, (3) improving connections between academics and industry, and (4) bringing about systemic and lasting changes in the way chemistry is taught.

2.6.2 The Green Chemistry commitment

The goals of the faculty involved in the *Green Chemistry Commitment* include uniting the academic community around a shared set of student learning objectives, providing a way to track progress of the community, and providing direction for outreach and advocacy. That is the reason why the author has chosen to incorporate the term "Community of Practice" into the description of the *Green Chemistry Commitment*

The section of the *Green Chemistry Commitment* document devoted to departments or institutions states that the department that commits itself to signing this document believes that *all* chemistry majors should be proficient in the following green chemistry competencies upon graduation:

Theory: Have a working knowledge of the twelve principles of green chemistry.

Toxicology: Have an understanding of the principles of toxicology, the molecular mechanisms of how chemicals affect human health and the environment, and the resources to identify and assess molecular hazards.

Laboratory Skills: Possess the ability to assess chemical products and processes and design greener alternatives when appropriate.

Application: Be prepared to serve society in their professional capacity as scientists and professionals through the articulation, evaluation, and employment of methods and chemicals that are benign for human health and the environment.

The document that indicates that a department is going to adopt the *Green Chemistry Commitment* requires the signature of both the chair of the chemistry department and at least one administrator at the level of dean, provost, or university president to confirm the notion that this is, in fact, a long-term commitment by the department to these goals.

The student learning objectives/competency related to toxicology in the *Green Chemistry Commitment* deserves special attention. Proponents of this competency should appreciate a point Anastas made in the interview published in *Nature*

> When I was getting my PhD in chemistry I was expected to translate technical articles from French and German to English ... But I can tell you I have never, ever, had to translate an article ... in all of my working life. And yet there was never any requirement that I needed to know the first thing about toxicity or the hazards of the tools of my trade – the chemicals. I never had to take a test that required me to understand the consequences of the molecules that I was introducing into the Universe or the ones that I was using on a daily basis. There's an absurdity there that needs to be addressed [14]

While discussing the Green Chemistry Commitment with faculty at various institutions, the author has noted that the challenge many of these individuals believe they face is finding resources to introduce toxicology. The author has found that using the text on *Laboratory Safety for Chemistry Students* [29] as part of an introduction to safety in a sophomore seminar course provides a basis for introducing some of the basic ideas of toxicology into the chemistry major curriculum. A conversation with executives at Chemical Abstracts Service about the use of SciFinder as a resource in this area suggested that information about toxicology might be available in this resource for perhaps 250,000 of the more than 100 million compounds in the database. It might also be noted that there is an ACS online short course entitled *Toxicology for Chemists* [30], a text entitled *Fundamental Toxicology For Chemists* published by the RSC [31], a toxicology data network (TOXNET), and handbooks on toxicology [32, 33]

2.7 The "triple bottom line" in academics

When examples of Green Chemistry are threaded into the fabric of the course, it doesn't take much time to convince students of the legitimacy of the "triple bottom line" within the context of practicing chemists working in industry. The author's experience has shown that they soon recognize that there are economic, social and environmental benefits when *someone else* adopts this mode of thinking. But that does not mean that they also appreciate that similar benefits can be achieved within their own activities as students.

One way of getting them to think about the implications of the Green Chemistry movement on academics is to give them examples from the article entitled "It's even cheaper being green," which appeared in the *Chemistry World* magazine distributed to RSC members [34]. That article showed how the investment of funds in the short term can have long-term returns in the form of recurring lower costs for the department. Andrea Sella, from University College London is quoted, for example, as noting: "… because energy is essentially invisible, most of us have little or no idea even of the order of magnitude of what we're using." The vice-chancellor at Cambridge, himself a chemist, was quoted as noting that chemistry labs are often found in older buildings that have been remodeled, "… so you end up with a complicated building that nobody really fully understands," adding: "Facilities managers have a complex job just trying to keep water flowing through the pipes and nowhere else, and the electricity in the wires and nowhere else." The *Chemistry World* article noted the existence of a not-for-profit organization – S-Lab – whose director was quoted as saying: "'Three to four years ago, most people didn't imagine that you could do things differently in chemistry labs … They thought it was all a given; that the designs of the ventilation systems, which account for the majority of lab energy use, were fixed; the way you operate is the way it's always been done and that's the way it will always be done. But people are beginning to wake up to the fact that there are alternatives."

In the same year that the Pollution Prevention Act was enacted by Congress, a volume entitled *Assessment in the Service of Instruction* [35] was published. In the preface, Shirley Malcom argued: that "... assessment should at least:

- be free of bias,
- reflect what is being taught and give us information to improve instruction for a class or to diagnose problems or to identify misconceptions of an individual,
- allow us to measure the effectiveness of a teacher or a curriculum,
- reflect what *should* be taught or at least what should be valued."

In the same volume, Lovitts and Champagne [36] stated what might be considered an "eternal verity" – something that will be true until the end of time – when they noted that: "It is generally conceded that what does not get assessed does not get taught."

If the Green Chemistry movement is something to be valued, it must also be incorporated into student assessment. But, it should be remembered that the idea that something should be assessed does not mean it has to appear on exams; it can also be incorporated into the assessment of lab activities associated with the course. A variety of questions can be envisioned, including: Given the difference between the properties of THF and 2-methylTHF, list five reasons why the use of 2-methylTHF would be consistent with a Green Chemistry philosophy? What aspects of the lab done this week do you believe were changed by your instructor when they tried to bring the Green Chemistry thought process into this experiment? In what ways might this week's experiment be improved by introducing changes that represent the philosophy of the Green Chemistry movement? In what ways could this week's experiment be made safer?

If one believes that assessment is best associated with exams, there are a lot of relatively simple open-ended questions available to you. Where is the idea of the "triple bottom line" exhibited in the following guiding principles: *Design less hazardous chemical syntheses, design safer chemicals and products*, and *use safer solvents and reaction conditions?* Or, in what ways is the guiding principle advocating chemists to *design chemicals and products to degrade after use* consistent with the efforts at your institution to implement a campus-wide commitment to the green environmental movement? Or, in what ways are the examples of Green Chemistry you've experienced in this course examples of efforts toward the goal of sustainable development?

As an example of the effect the guiding principles of Green Chemistry has had on the author's life in academics, think about the factors that needed to be considered when he first joined the faculty at Purdue, 40 years ago. At that time, Purdue had more than 5000 students taking General Chemistry each fall semester. Imagine the aspects of size and scale he faced when trying to introduce new experiments into just one of the courses in this program, which enrolled roughly 2500 students. Any experiment added to the curriculum had to be compatible with the equipment available in each of the thousands of laboratory drawers. The cost of purchasing chemicals was an important constraint because literally thousands of students were doing the same experiment each week. Safety, of course, was another important factor influencing the design of new experiments.

Now, try to imagine the constraining forces he would face if he was trying to do the same thing today. It should be easy to appreciate that the cost of disposal of waste products generated in any experiment introduced into the curriculum may have become a more important factor than the cost of purchasing the chemicals in the first place.

2.8 Reflections on the evolution of Green Chemistry in academics

Ten years ago, The National Research Council [37] published a report entitled *Exploring Opportunities in Green Chemistry and Engineering Education* that was based on a two-day workshop organized by the Chemical Sciences Roundtable. Participants at this workshop correctly noted the existence of "green islands" corresponding to "relatively small pockets of activity in green chemistry education." But they also correctly noted that "... sustainability is the single most important challenge for our civilization for at least the next 100 years."

Recently, the ACS GCI® completed a review of the status of green chemistry education in the United States [38]. The author of this chapter concluded from reading this report that additional "green islands" are gradually appearing and, for the first time, these islands include a representative sample of research-intensive institutions. Consider, for example, institutions that are members of the Association of American Universities (AAU), which includes universities "... on the leading edge of innovation, scholarship, and solutions that contribute to the nation's economy, security, and well-being." This group of 60 research-intensive institutions "... award nearly one-half of all U.S. doctoral degrees and 55 percent of those in the sciences and engineering" [39]. At the time this chapter was written, 25 % of the members of the AAU had a green chemistry academic program, including: Berkeley, Cal Tech, Carnegie Mellon, Florida, Georgia Tech, Indiana, Iowa State, Michigan State, Michigan, Ohio State, Oregon, Pittsburgh, Stony Brook (SUNY), Washington, and Yale.

The ACS GCI® report noted the widespread commitment at the institutional level among colleges and universities to the philosophy of the green environment movement. What the author finds frustrating is that this campus-wide program often exists without any involvement of the Chemistry Department.

This report also noted that, in spite of the dissemination of resources available to help instructors introduce green chemistry into their courses, most of the green chemistry courses being offered are introductory courses rather than upper-level courses designed to meet the needs of students enrolled in STEM majors. The ACS

GCI® report also noted that the majority of NSF funds (79 %) devoted to research on green chemistry and engineering were given to large, research-intensive institutions, and yet it was the smaller schools who were more likely to publically promote green chemistry initiatives.

In discussions with colleagues in chemical education from a wide variety of institutions and with staff at the ACS GCI®, the author has reached several conclusions.

- When Green Chemistry is introduced it is usually the result of a "bottom-up" rather than "top-down" implementation; it is virtually always introduced because of the beliefs or values of one or more faculty at the bottom of the organization chart, rather than because it was mandated by a department head or dean who occupies the top of chart.
- When Green Chemistry is introduced it is done so passionately, by instructors whose commitment to the principles of Green Chemistry and Sustainability are absolutely remarkable.
- Green Chemistry is often endorsed with equal passion by undergraduate chemistry majors who work with these instructors.
- Green Chemistry is most often introduced in a way that is the most convenient; by incorporating it into existing courses, rather than by creating a separate course, although separate, upper-level courses on green chemistry do exist.
- Separate, upper-level Green Chemistry courses are unusually "fragile" or "volatile." The author is familiar with a number of instances where an upper-level course was created, but soon vanished because well-intentioned instructors who created the Green Chemistry course were pulled out of these courses to staff more traditional courses.
- Although there are examples of institutions where a Green Chemistry commitment has been integrated throughout the undergraduate curriculum, they are still rare, and they are most likely to occur in relatively small departments where it is significantly easier to reach the consensus necessary to do this.

The author has also noted that the twenty-first century is following a pattern that characterized curriculum development/reform projects throughout the second-half of the previous century. The first characteristic of this pattern, which has appeared over and over again for at least 60 years, can be appreciated by noting that far more time, effort, and resources go into the development of curriculum materials than into dissemination of these materials, with the notable exception of the Green Chemistry group at the University of Oregon who have done an extraordinary job in getting materials distributed among people who would be likely to use them [40].

The second characteristic of curriculum development/reform projects can be seen by noting that, however small the amount of time, effort and resources devoted to dissemination of the curriculum materials, it is still an order of magnitude (or more!) larger than the time, effort and resources devoted to the evaluation of the effect of the new curriculum materials on students or their instructors. Periodically, one encounters papers whose authors are honest enough to note that no formal, large-scale evaluation program was carried out because all of the funding was tied to the development and dissemination of the curriculum materials [41].

2.9 Why there might be a "quadruple bottom line" in academics

The title of this chapter was taken from a paper presented by the author in a symposium at the 2016 Biennial Conference on Chemical Education. It builds on the idea of the triple bottom line that was one of the take-home messages from the author's interactions with members of both the Pharmaceutical Roundtable and the Chemical Manufacturers Roundtable at the first ACS GCI® conference he attended. He has found that undergraduates readily accept the idea that Green Chemistry can have strong, positive effects within the domains of economic, social and environmental benefits. But these factors, alone, cannot explain why so many students respond so enthusiastically to the Green Chemistry movement when the examples cited by their instructors seldom, if ever, seem to have a direct impact on what happens in the students' daily lives. To explain that effect, we need to digress to understand the fourth element in the quadruple bottom line: *relevance*.

At first glance, most readers of this chapter will assume that the term *relevance* needs no further explanation. If they happen to look at the definition of *relevance* on the Merriam-Webster website they might even note the apparent validity of the example used there: "giving *relevance* to college courses." But please forgive the author as he asks for a moment to expand on the history of the call for "relevance" in our courses and then describes a recently proposed model that clarifies some of the inconsistencies in the way the term is used when referring to the chemistry classroom.

The chapter entitled "Understanding the change toward a greener chemistry by those who do chemistry and those who teach chemistry" [11] to which the author has periodically referred was written for his colleagues in "science education." It is an example of something he has done repeatedly throughout his career – writing about chemistry for those doing educational research or writing about the results of research in education for chemists. The chapter appeared in a book [42] that built on a paper that probed the question: What does it mean to ask teachers to make their course "more relevant" to students? [43].

2.10 A new way of looking at "relevance"

It has been 45 years since Gallagher [44] proposed the creation of "STS" courses that would show how science is relevant to everyday life by integrating science, technology and society. By 1982, the National Association of Science Teachers began a report on STS courses by noting: "The goal of science education during the 1980s is to develop scientifically literate individuals who understand how science, technology and society influence one another and who are able to make use of this knowledge in their everyday decision making. This individual both appreciates the value of science and technology in society and understands their limitations" [45].

With support from NSF, the ACS took a step toward an STS course by developing a "context-based" high-school chemistry course known as *Chemistry in the Community* intended to "provoke student interest and involvement in chemistry … by embedding chemistry within society, where chemistry daily impacts human lives, rather than metaphorically confining chemistry to laboratory flasks … and brown bottles" [46]. In other words, to make the high-school course more *relevant*. The success of *Chem Comm* led the Society to create a one-semester, college-level text for non-science majors known as *Chemistry in Context* [47]. More or less the same factors led to the creation of "context-based" chemistry courses in other countries at more or less the same time [41, 48, 49].

To illustrate the problem with the idea of making chemistry courses *more relevant*, let's return to the period when so much activity surrounding the development of "context-based" chemistry courses occurred and note that Newton [50] began a paper on "relevance and science education" by noting: "… science teachers are increasingly exhorted to make their teaching relevant but, in general, the notion of relevance in science education seems fraught with inconsistency, obscurity and ambiguity." Newton noted that the term *relevance* is sometimes used within a context that allows students to relate science to their daily lives; at times it seems to imply connecting the course content to their prior experience or existing ideas students bring to the course; whereas at still other times it seems to imply that the material being taught should be related to the world they will experience as adults. Newton concluded the introductory section of his paper as follows: "The notion of relevance is not a simple one. It seems at the least unhelpful and at worst counterproductive to urge a teacher to be relevant in terms which are abstract and diffuse. It might be useful if some aspects of the notion of relevance were to be clarified." Perhaps it should not be surprising that debates about *relevance* in science education have been going on for more than 40 years [51].

Recently, a model developed by Stucky et al.,has been proposed that might help us understand that there are three dimensions within which *relevance* can be viewed [43]. Each of these dimensions can be viewed as a plane defined by the same pair of axes. One axis reflects the continuum between two points in time, now and the future. The other can be viewed in terms of a perspective that builds on *Self-Determination Theory* [52, 53], which distinguishes between different types of motivation by noting: "The most basic distinction is between intrinsic motivation, which refers to doing something because it is inherently interesting or enjoyable, and extrinsic motivation, which refers to doing something because it leads to a separable outcome" [53].

The first dimension of relevance can be viewed as the *individual dimension*. Chemistry, in general, or Green Chemistry, in specific, can be thought of in terms of a student's innate interest or curiosity in one corner of a plane corresponding to both the "present" and an "intrinsic" motivation. Another corner, corresponding to the "present" and "extrinsic" motivation might be thought of in terms of the student's goal of getting a good grade. As one moves toward the corner defined by "future" and "intrinsic," one might find the idea of material that might be associated with an individual's developing skills to help them succeed in life. And the combination of "future" and "extrinsic" might be associated with acting responsibly as individuals in their future lives.

The *societal dimension* of relevance can be viewed using the same axes; present versus future and intrinsic versus extrinsic. The corner associated with the present and an intrinsic point of view might be thought of in terms of the students' efforts to find their own place in society, whereas future and intrinsic could be viewed as promoting one's own interests in the societal discourse they will encounter. The present and extrinsic corner might be associated with learning how to behave in society, whereas the future and extrinsic corner might be seen as behaving as a responsible, adult citizen within the context of society, as a whole.

The *vocational dimension* can be seen as a continuum along the set of points characterized by an intrinsic motivation as one views the present emphasis on developing the skill set that orients the student toward potential careers moving toward the goal of obtaining a job one can value. Moving along the extrinsic side of the diagram one can envision a present focus on doing well enough to qualify for subsequent courses at this point in time toward the important future goal of contributing to society's economic growth.

The author has found the model developed by Stuckey et al. [43] useful as he thinks about how to bring the final element of the quadruple bottom line into his classroom, regardless of which chemistry course he is teaching that semester. Hopefully, others will find it useful as well because all three dimensions of *relevance* are important as instructors who are passionate about the role that Green Chemistry can play interact with students – regardless of their major – who enroll in our courses as a step toward both their future careers and their participation as a scientifically literate citizen.

References

[1] Wenger E. Communities of practice: learning, meaning, and identity. Cambridge, UK: Cambridge University Press, 1988.

[2] Macklin AS. Theoretical frameworks for research in chemistry/science education. In: Bodner GM and Orgill M, editors. Communities of Practice Upper Saddle River, NJ: Prentice Hall, 2007:195–217.

[3] Surowiecki J. The wisdom of crowds: why the many are smarter than the few and how collective wisdom shapes business, economies, societies and nations. New York, NY: Doubleday, 2004.

[4] Rosebery AS, Warren B, Conant FR. Appropriating scientific discourse: findings from language minority classrooms. J Learn Sci. 1992;2:61–94.

[5] Brown JS, Collins A, Duguid P. Situated cognition and the culture of learning. Educ Res. 1989;18:32–42.

[6] Green Chemistry. Aug 2014 Available at: http://www2.epa.gov/green-chemistry/. Accessed27Aug2014.

[7] International Institute for Sustainable Development. Aug 2014 Available at: http://www.iisd.org/sd/. Accessed27Aug2014.

[8] World Commission on Environment and Development (WCED). Our common future. Oxford, UK: Oxford University Press, 1987:43.

[9] Green Chemistry. Aug 2014 Available at: http://www.rsc.org/ScienceAndTechnology/Policy/EHSC/EHSCnotesonGreenChemistry.asp/. Accessed27Aug2014.

[10] Beyond Benign: Green Chemistry Education. Aug 2014 Available at: http://www.beyondbenign.org/. Accessed27Aug2014.

[11] Bodner GM. Understanding the change toward a greener chemistry by those who do chemistry and those who teach chemistry. In: Eilks I, Hofstein A, editors. Relevant chemistry education – from theory to practice. Rotterdam: Sense Publishers, 2015: 263–284.

[12] Pollution Prevention: Definitions. Aug 2014 Available at: http://www.epa.gov/p2/pubs/p2policy/definitions.htm/. Accessed27Aug2014.

[13] Green Chemistry Definition. Aug 2014 Available at: https://www.acs.org/content/acs/en/greenchemistry/what-is-green-chemistry/definition.html/. Accessed27Aug2014.

[14] Q&A: Paul Anastas. Aug 2014 Available at: http://www.nature.com/news/2011/110105/full/469018a/box/2.html/. Accessed27Aug2014.

[15] About Green Chemistry. Aug 2014 Available at: http://www.warnerbabcock.com/green_chemistry/about_green_chemistry.asp/. Accessed27Aug2014.

[16] Anastas PA, Warner JC. Green chemistry: theory and practice. Oxford, UK: Oxford University Press, 2000.

[17] Basics of Green Chemistry. Aug 2014 Available at: https://www.epa.gov/greenchemistry/basics-green-chemistry#twelv/. Accessed27Aug2014.

[18] Anastas PT, Zimmerman JB. Design through the 12 principles of green engineering. Environ Sci Tech. 2003;37 94A-101A.

[19] Daly S, Adams R, Bodner GM. What does it mean to design? A qualitative investigation of design professionals' experiences. J Eng Educ. 2012;101(2):187–219.

[20] History of the Green Chemistry Institute. Aug 2014 Available at: http://www.acs.org/content/acs/en/greenchemistry/about/history.html/. Accessed27Aug2014.

[21] What is Green Chemistry? Aug 2014 Available at: https://www.acs.org/content/acs/en/greenchemistry/what-is-green-chemistry.html/. Accessed27Aug2014.

[22] Vorhees K, Hutchison JE. Green Chemistry Education Roadmap charts the path ahead. Chem Eng News. 2015;93(38):46.

[23] Synthesis of ibuprofen. Aug 2014 Available at: http://www.epa.ohio.gov/portals/41/p2/ibuprofencasestudy.pdf/. Accessed: 27Aug2014.

[24] Sildenafil Citrate: Green Chemistry. Aug 2014 Available at: http://www.tes.co.uk/teaching-resource/Sildenafil-Citrate-Green-Chemistry-6397172/. Accessed27Aug2014.

[25] Greener Alternatives. Aug 2014 Available at: http://www.sigmaaldrich.com/chemistry/greener-alternatives.html/. Accessed27Aug2014.

[26] Mowry J. Way past cool. New York, NY: Farrar Straus Giroux, 1992.

[27] Beyond Benign: Green Chemistry Education. Aug 2014 Available at: http://www.beyondbenign.us/home/about/about.html. Accessed27Aug2014.

[28] Green Chemistry Commitment. Aug 2014 Available at: http://www.greenchemistrycommitment.org/. Accessed27Aug2014.

[29] Hill RH, Finster D. Laboratory safety for chemistry students, 2nd ed New York: John Wiley & Sons, 2016.

[30] ACS. Toxicology for chemists online short course. Aug 2014 no date. Available at: http://proed.acs.org/course-catalog/courses/toxicology-for-chemists-online-short-course/. Accessed27Aug2014.
[31] Duffus JH, Worth HG. Fundamental toxicology for chemists. Cambridge, UK: Royal Society of Chemistry, 1996.
[32] Derelanko MJ, Auletta CS. Handbook of toxicology, 3rd ed Boca Raton, Fl: CRC Press, 2014.
[33] Murray L, Little M, Pascu O, Hoggett K. Toxicology handbook, 3rd ed Australia: Elsevier, 2015.
[34] Broadwirth P. It's even cheaper being green. Chem World-UK. 2014 May;2014:50–53.
[35] Champagne AB, Lovitts BE, Calinger B. Assessment in the service of instruction: papers from the 1990 AAAS forum for school science. Washington, DC: American Association for the Advancement of Science, 1990.
[36] Lovitts BE, Champagne AB, Champagne AB, Lovitts BE, Calinger B. Assessment in the service of instruction: papers from the 1990 AAAS forum for school science. Washington, DC: American Association for the Advancement of Science, 1990:1–13.
[37] National Research Council. Exploring opportunities in green chemistry and engineering education: a workshop summary to the chemical sciences roundtable. Washington, DC: The National Academies Press, 2007.
[38] Constable D. Personal communication 2014.
[39] Association of American Universities. August 2014 Available at: https://www.aau.edu/. Accessed27August2014.
[40] University of Oregon Green Chemistry Program. Aug 2014 Available at: http://greenchem.uoregon.edu/. Accessed27Aug2014.
[41] Bennett J, Lubben F. Context-based chemistry: the Salters approach. Intern J Sci Educ. 2006;28:999–1015.
[42] Eilks I, Hofstein A. Relevant chemistry education – from theory to practice. Rotterdam: Sense Publishers, 2015.
[43] Stuckey M, Hofstein A, Mamlock-Naaman R, Eilks I. The meaning of 'relevance' in science education and its implications for the science curriculum. Stud Sci Educ. 2013;49:1–34.
[44] Gallagher JJ. A broader base for science teaching. Sci Educ. 1971;55(3):329–338.
[45] National Science Teachers Association. Science-technology-society: science education for the 1980's. Washington, DC: National Science Teachers Association, 1982.
[46] Heikkinen HW. To form a favorable idea of chemistry 1. J Chem Educ. 2010;87:680–684.
[47] Schwartz AT, Bunce DM, Silberman RG, Stanitski CL, Stration WJ, Zipp AP. "Chemistry in Context": weaving the web. J Chem Educ. 1994;71:1041–1044.
[48] Hofstein A, Kesner M. Industrial chemistry and school chemistry: making chemistry studies more relevant. Intern J Sci Educ. 2006;28:1017–1039.
[49] Parchmann I, Gräsel C, Baer A, Nentwig P, Demuth R, Ralle B. "Chemie im Kontext": A symbiotic implementation of a context-based teaching and learning approach. Intern J Sci Educ. 2006;28:1041–1062.
[50] Newton DP. Relevance and science education. Educ Phil Theor. 1988;20(2):7–12.
[51] Fensham PJ. Increasing the relevance of science and technology education for all students in the 21st century. Sci Educ Intern. 2004;15:7–26.
[52] Deci EL, Ryan RM. Intrinsic motivation and self-determination in human behavior. New York: Plenum, 1985.
[53] Ryan RM, Deci EL. Intrinsic and extrinsic motivations: classic definitions and new directions. Contemp Educ Psychol. 2000;25:54–67.

Larry Kolopajlo

3 Green chemistry education in the Middle East

Abstract: The Middle East once dominated the age of alchemy, and today it is experiencing a resurgence by transforming the age of petroleum chemicals into a greener science through *Estidama*. This green conversion is taking place through green chemical research and education. This report examines and reviews the understudied subject of green chemical education in the Middle East through the lens of context and history.

Keywords: Middle East, green chemistry education, sustainable chemical education

3.1 Introduction

Why write a paper on green chemical education (GCE) in the Middle East? The answer to this question is not so simple because the Middle East (ME) stretches across a vast, diverse and politically complex region. One reason would be that since the ME is underserved, and lagging behind the West in having well-rounded GC production and regulatory policies, then understanding how GCE evolves in the ME will allow for both corrective action and future planning. Another reason for such a paper is that no general review of the subject exists, and a paper on GCE-ME would complement published books on Green Chemistry (GC) in: Africa [1], Latin America [2], Russia [3]. However, the author of this paper did not conceptualize of writing about green chemistry education (GCE) in the Middle East as a way to complete another volume in a series. Instead, the author became interested in GC-ME while working with middle eastern students at Eastern Michigan University, which has a diverse student body. A third reason for this study is that as a global hub for petroleum and petrochemicals, there is a need to address GCE in the ME, but within its unique set of social and environmental contexts. Furthermore, as the ME makes strides toward a knowledge-based economy, it becomes more ready to accept and implement the philosophy of GCE in order to improve quality of life.

Green chemistry (GC) research is already well established in the ME where researchers have been prolific generators of thousands of published GC papers on a wide variety of subjects. However, unlike in the United States, no formal GC academic programs confer GC degrees. GCE therefore occurs through a number of ancillary ways involving research and collaboration in academic chemistry and chemical engineering programs, interdisciplinary programs, seminars and workshops at conferences, professional societies, government, industry, outreach, and social media. In this report, each of these methods of disseminating GCE will be discussed and elaborated.

https://doi.org/10.1515/9783110445923-003

Moreover, this paper will demonstrate the need for GCE by examining its ME context with respect to geography and culture, and emerging trends in population dynamics, waste generation, water, environmental pollution, and climate change through a cogent analysis of widely available information. The status and evolution of GCE-ME will then be reviewed. Specific studies illustrating important GCE-ME contributions to the field will be highlighted on a country by country basis, or through delineating work supported by specific organizations and institutions. It will be seen that GCE-ME is developing incrementally, often first in interdisciplinary fields such as green pharmaceutical chemistry or green petroleum chemistry, and that it sometimes germinates through outreach, professional societies and conferences, or through the work of government.

3.2 Background

3.2.1 Estidama

One way to understand the ME context of GC is to appreciate the Arabic word for sustainability, Estidama, and understand its meaning. The ME context of Estidama includes the ability to build thriving civilizations in regions where nature offers formidable challenges. The concept is thus familiar to the ME and suggests a minimalist philosophy of living, a concept akin to GC.

In the ME, sustainability has preceded GC. Dubai is a recognized world leader in sustainability and its gleaming metropolis is renowned for its sustainable infrastructure. At the individual level, at the Canadian University of Dubai, Professor Hoshiar Nooraddin [4] has been very active in promoting education for sustainable development. Saudi Arabia has also incorporated sustainability for society, economy, and future needs through sound business practices and social responsibility, sometimes referred to as "People, Planet and Profit" [5]. So there is a need for cogent governmental strategies to integrate and promote synergy between industry and academia. The ME needs to transition from where it is now to a more sustainable society, and this is being done with education for sustainable development (ESD).

Although some countries have made great strides in achieving sustainability, fewer have devoted as much effort into GC. The overall strategy first deploys sustainability, and then sustainable chemistry before GC. In contrast to GC which is more pure chemistry, sustainable chemistry has a more applied context, and includes social responsibility; in a nutshell, it is chemistry and society. Sustainable chemistry thus promotes economic, environmental, and social responsibility as a new paradigm. Hence sustainable chemistry takes place at the intersection of science, technology, and culture, and embodies a interdisciplinary context. Therefore, the transformation from sustainability to GC is taking place incrementally rather than in broad swaths, and GCE, although on the periphery of the sustainable movement in the ME, is still playing an important role by pressing for change. In this paper, those discrete efforts to promote GCE will be examined.

3.2.2 Geography

Technically speaking, a textbook definition of ME geography would denote that it encompasses 17 countries from Turkey on the west to Afghanistan on the east, traversing both northern Africa and western Asia. But Middle Eastern culture and religion also stretch across all of Saharan Africa from where Morocco meets the Atlantic Ocean, to Egypt on the Gulf of Suez. Moreover, the state of Israel, not being a member of the European Union, is also a Middle Eastern country. Therefore, in this paper, the Middle East designation will be more inclusive than what a textbook definition might offer, and also, for example, include those countries located in Saharan Africa, expanding the ME data base to many more countries, and about 250 million people [6].

3.2.3 Population

By 2050, several factors are predicted to transect creating an apocalyptic scenario. Overpopulation is forecast to stretch the capacity of the earth to support 2.5 billion more people than its current population of 7.3 billion people [7], half of whom may live in a totally different and perhaps unkinder climate [8]. Although half of the population increase is projected [9] to come from Africa, India, Indonesia, and the USA, the ME Muslim population is predicted to increase by about 37 % by 2030. Egypt, the most populous ME country, is projected to increase its population to about 114 million by 2050 [10].

One goal of climate treaties being negotiated is to decarbonize the world by 2050 through decreasing petroleum usage, and by widespread implementation of mitigating emissions through carbon capture and carbon storage technologies. But as the global population grows, game changing technologies will be required to ensure a sustainable society, and to correct problems associated with water, waste, pollution, natural resources, and energy. These problems will be solved through collaborations between industry, government, and education. ESD and GCE will need to play an important role in educational curricula.

3.2.4 Pollution and waste

With increasing population, more waste and pollution are generated, and more renewable and nonrenewable resources are required to sustain society. Although the Blacksmith Institute [11] does not rank any middle eastern country as owning one of the world's worst pollution problems, major petroleum producing countries periodically cause oil leaks, generate waste products, and damage the aquatic environment through oil spills. Thankfully, in the ME there haven't been any disasters on the order of the Exxon Valdez (1984), or Bhopal (1984), demonstrating that ME countries have learned from some of history's greatest chemical tragedies.

However, one pressing problem in the ME is air pollution. For example, according to an article published in Environmental Science and Technology [12], Mecca, Saudi Arabia, which has a normal population of about 2 million and swells to 6 million during pilgrimages, suffers from heavy air pollution.

Another extensive problem facing the ME is waste. Middle Eastern countries with large populations have generated so much per capita waste that it has overwhelmed cities. Some countries have been forced to adopt recycling and waste-to-energy conversion measures. In Iraq, for example, solid waste production is commonly disposed of in unregulated landfills because of a depleted infrastructure, and as a result, these landfills have sometimes caused fires, water pollution, and large greenhouse gas emissions [13]. In 2007, the National Solid Waste Management Plan (NSWMP) for Iraq was developed to promote sustainable development [14].

3.2.5 Water

According to a UN Human Development Report titled: *Water Scarcity* Challenges *in the Middle East and North Africa* [15] published in 2006, water has been a scarce commodity in the ME and North Africa since the 1970s. The current efficiency of water usage is only 40 %, and with a rising population and increased urbanization, water must be managed more effectively [16]. Currently, irrigated agriculture consumes most of the water, and much of that is lost from evaporation.

All renewable freshwater resources are being consumed in Saudi Arabia and its neighbors: Bahrain, Kuwait, Oman, Qatar, the UAE, and Yemen. Moreover, the same can be said of Israel, Jordan, Gaza, and the West Bank. Furthermore, these countries suffer from poor water quality: Algeria, Egypt, Iraq, Iran, Lebanon, Morocco, Syria, and Tunisia. In 2008, a severe water shortage caused the Jordanian government to enact an emergency strategy to deal with demand among its 5.7 million population and hundreds of thousands of refugees [17]. In Egypt, raw sewage contaminates the water ways which serve as makeshift garbage dumps. Egypt has also voiced concern over Africa's immense Grand Millennium Dam, a hydroelectric dam under construction on the Nile in Ethiopia [18]. The problem of water scarcity and pollution has also advanced a new concept, "Eco-peace" that has spurred cooperation between Israel and Jordan to save the Jordan River [19].

In Saudi Arabia, Riyadh is supplied with desalinated water that is pumped from the Persian Gulf nearly 500 km away. Water is also brought in as needed using trucks that patrol the streets. The grand Al Ahsa Oasis supports an enormous agricultural industry involving dates and rice, but one of its biggest problems is that water is being pumped from the ground at a faster rate than it can be replenished [20]. Moreover, pesticides and fertilizers are over-used and are leaching into the ground water [21]. To solve these problems, groundbreaking agricultural practices such as drip irrigation [22] and natural pesticides, as invented in Israel, must be substituted for more harmful practices. Moreover, extracting drinking water from the air may become necessary using technology like that invented by the Water-Gen [23] company in Israel. Abdallah El Maaroufi suggests a three-part strategy to resolve environmental problems, consisting of partnerships between countries, improving resource management, and strengthening participating institutions [24].

3.2.6 Economics

Within the expansive area of the ME are both rich and poor nations. Israel, Qatar, Kuwait and the United Arab Emirates are wealthy, while Saudi Arabia, Bahrain, and Oman have good incomes [25]. For the remaining countries, income is a challenge for many citizens. However, as the ME becomes more industrialized and independent, GC may provide jobs. For example, according to Pike Research [26], the global GC market will reach $5.3 trillion by 2020, and the Middle East GC Market will save the industry $65 billion by 2020 [27].

3.2.7 Organizations

Many organizations are actively promoting GC, GCE, and Education for Sustainable Development (ESD) in the ME, among them being the powerful UN (United Nations), the International Union for Pure and Applied Chemistry (IUPAC), and some regional scientific entities. In this section and subsequent ones, the contributions to GCE from such organizations will be described in more detail. Although the UN has published much about sustainable education, it has done little regarding GC and sustainable chemistry (SC). However, the UN acknowledges and recognizes the importance of GC and what must be done to institute it. For example, during the United Nation's Decade of Education for Sustainable Development from 2005–2014, supported by the American Chemical Society, all member countries educated their citizens on the importance of sustainability [28]. Sustainability came first, and GCE will come later.

3.2.8 Education

In order to understand how GCE fits into the ME context, one must understand something about ME educational systems. The United Nations publishes a Human Development Index (HDI) every year, which includes an Education Index [29]. Many ME countries are ranked as having high Education Indices. The 2013 report [30], for example, showed that Israel had the highest Education Index at 0.85, while Saudi

Arabia was at 0.723, with Norway having the world's highest value at 0.91. Thus, generally speaking, the ME possesses the educational system needed to support a GCE curriculum.

3.2.9 Academia

Although GCE has not been a priority in the ME, both industrial and academic researchers are very active in the field of GCE and have disseminated it in a variety of creative ways, sometimes using social media, such as ResearchGate, YouTube, and Slide Share. Their social postings may seed GCE as a GC social movement.

The position taken in this paper is that if researchers are in a ME academic institution doing GC, then GCE is involved because the training of future GC workers counts as GCE. Like in the U.S., in the ME, academic institutions are offering GCE, but indirectly through research. One can argue that students are picking up GCE pedagogy through a secondary channel. They are using the pedagogy of the field, but toward a different end, that involving research. Eventually, as more workers are trained in interdisciplinary fields, even in Education or the Social and Political Sciences, more GCE will evolve.

3.3 GCE contributions in the middle east

3.3.1 Types of green chemical educators

There appear to be three classes of ME green chemical educators, all working at academic institutions: Type I): trained research chemists who may for example, specialize in green organic chemistry, Type II): non-chemists working in an interdisciplinary field, and Type III): trained GC educators of which there are currently few in the ME, partly because there are no formal GC academic programs that confer academic degrees. In the next sections, contributions to Middle Eastern GCE will be reviewed, demonstrating how it has evolved.

3.3.2 Organizations

Now that the context of GC in the ME has been examined with respect to economics, education, and problems involving natural resources, GCE specifics can be addressed. In this section will be surveyed those UN supported ME-GCE projects for selected countries that surround part of the Mediterranean basin. However, admittedly, Lebanon, Libya, Israel, and Turkey are not covered in this section. Moreover, although there are a lot of acronyms in this section, the reader should focus on the main idea, that is: how and what kind of GCE is being promoted in the ME. Many of the GC programs described in this section involve UNESCO (the United Nations

Educational, Scientific, and Cultural Organization), which has long been involved in promoting sustainability, and more recently, GC. Other programs described in this section have been run jointly with the European Union (EU), NATO, and IUPAC, and as a result of these collaborations, a number of worthwhile programs involving GCE in the ME have been instituted.

This section will begin with the UNITWIN (University Twinning and Network Program) of UNESCO founded in 1992, which promoted international inter-university cooperation and networking to enhance institutional capacities through knowledge sharing, shared governance, and collaborative work [31]. Certain programs are promoting GC to jointly meet economic and environmental needs through collaborations between industry, government, and academic institutions.

One program that promoted GC and GCE was instituted in 2005 when G8 Ministers for Research founded a research and training network on green sustainable chemistry called the International Green Network (IGN) whose hub was located in Venice [32]. Some IGN objectives were to sponsor, coordinate, and provide information for GC scientific collaborations, and to provide training for young scientists.

Another very important and effective program was initiated in 1993, when INCA (The Interuniversity Consortium, Chemistry for the Environment) was founded to promote environmental research among its thirty-three Italian member institutions [33]. Its host office is in Venice. INCA has been supported by UNESCO and NATO-ASI (Advanced Sciences Institute). Some of the work currently being done by INCA now includes sustainability and GC, and as will be later shown, INCA has evolved to serve the ME.

The INCA Summer School on GC served ME countries, especially those in the Mediterranean basin. It was held from 1998 to 2005, often in Venice, Italy, and after 2005, it was then supported under a NATO-ASI banner [34]. It was originally funded through a grant from the European Commission's IV Framework Programme (FP) Training and Mobility of Researchers (TMR) program, and continued within FP-TMR as part of an improved program, but it has also received funding through INCA. For example, the 7th summer school held on Servolo Island in 2005 received UNESCO support [35].

With respect to GCE, one highlight of the 2006 NATO-ASI summer school (IX) was a workshop titled "New Organic Chemistry Reactions and Methodologies for Green Production." This workshop was supported through a program that was co-directed by: Prof. Pietro Tundo, (Consorzio, INCA) in Marghera, Italy; and Prof. Ahmed Tawfic, of Suez Canal University in Ismailia, Egypt [36]. Although it isn't obvious, the 2008 summer school program objectives involved GCE because the published program indicated that "The teaching will be divided in basic themes (Atom economy, Industrial Processes, Alternative Solvents, New Feedstocks and Products, New Reactions and New Synthetic Methods), and special topics selected according to the availability of the teachers. In addition, topics related to current research in GC will be addressed with the aim to familiarize the students with the strategies behind the planning and designing of efficient and "greener" synthetic routes." [37]. Although the summer school programs just described emphasized GC research over GCE, at these events there has been a substantial amount of teaching and training, some of it involving the ME. For example, the 1998 conference featured Paul T. Anastas while the 1999 program featured Guy J. Martens of Belgium-Solvay [38]. These programs emphasize educating younger chemists on green research in chemistry, so that they will become part of the network of European green chemists.

In 2006, The Mediterranean Basin Green Chemistry Network (MEGREC) was founded with UNESCO-UNITWIN funding in order to catalyze the transformation to green and sustainable chemistry and technologies in the Mediterranean basin through conferences, seminars, workshops, and collaboration [39]. Its home institution was the Ca' Foscari University of Venice. One of MEGREC's stated goals was to increase the effectiveness of teaching and training activities in GC by integrating education and fundamental research. MEGREC's ME members institutions are: Mentouri University of Constantine in Algeria, Suez Canal University in Egypt, Sidi Mohamed Ben Abdellah University in Morocco, and the University of Gabès in Tunisa, and other non-ME institutions in Greece, Spain, and Serbia as well.

Building on the success of MEGREC, the Sustainable Middle East Development Initiative (SUSMEDI) project promoting GC centers of excellence was founded in 2012 [40]. For example, Morocco became a green catalysis hub, and Cairo focused on energy research while Tunisia centered on decontamination; Algeria became a center for environmental analysis. SUSMEDI was designed to have far reaching GCE effects, as for example, was shown by the development of a K-12 GCE outreach curriculum that (a) developed 12 lesson plans, one for each of the principles of GC and (b) developed a GCE outreach program for 5th graders.

Established in 1990, TEMPUS Joint European Projects [41] supported a curriculum reform project on sustainable environmental development. Later, TEMPUS was sponsored by the European community, and Italian Interuniversity Consortium that also runs INCA, Chemistry for the Environment. There were several phases of TEMPUS programs offered: a) TEMPUS III was carried out between 2000 and 2006 while b): TEMPUS IV between 2007 and 2013. In 2007, the TEMPUS program was initiated by the European Union (EU) and partner countries to reform and modernize higher education through a consortium of partner universities and university associations in neighboring countries such as those that surround the Mediterranean basin. TEMPUS is managed by the Education, Audiovisual, and Culture Executive Agency (EACEA).

In 2012, TEMPUS emphasized sustainable development and environmental monitoring of the Mediterranean region (SUSMED). All member countries took part. Some examples of ME TEMPUS projects are described below [42].

1. In 2002, project 30031-2002 in chemistry and biochemistry supported a curricular project on upgrading sustainable environment development with regard to university courses on environmental sustainability and GC, and to establish a consulting service for industry. The grant holder for this project was Tundo Pietro of INCA, and the coordinator was Tawfic Ahmed Mohamed – Suez Canal University Environmental Impact Assessment Unit – Ismailia, Egypt.
2. Project 32005-2004 supported a Master of Science Course in Applied Environmental Geosciences and Water Resources Management at Assiut University, Assiut, Egypt.
3. Project 30057-2002 was carried out at the Jordan University of Science and Technology supporting the creation of a education center on renewable energy.
4. In Lebanon, project 33056-2005 supported a master's program on sustainable energy at the American University of Beruit.
5. Project 31141-2003 supported a Master's program on environmental sustainability with the Palestinian Authority.
6. Project 31109-2003 assisted curriculum development and faculty training on renewable energy programs at Damascus University.
7. Project 33048-2005 promoted laboratories and training in solar energy in Syria, and in Algeria, while Project 31062-2003 addressed a Masters Degree in Environmental sustainability and pollution modeling.
8. TEMPUS ME projects in 2003 involved: Algeria, Egypt, Jordan, Lebanon, Morocco, the Palestinian Authority, Syria, and Tunisa.

3.3.3 Egypt

This section will highlight GC educators in Egypt, where several leaders work at these institutions: University of Cairo, Suez Canal University, and Ain Shams University. In a published paper titled "Sustainable Chemistry," Mohamed Tawfic Ahmed, a Suez Canal faculty member in Agriculture, provided a vision for Egypt's GC education programs by describing its challenges in the context of the Egyptian experience [43]. He noted the need for safe, environmentally friendly products, and the need for a chemical strategy that will provide a sustainable future for its citizens. He explained that Egypt has not yet embraced preventive measures to stop pollution, and needs to develop new, clean energy resources such as solar, and to use sustainable chemistry to develop a sustainable future by reducing waste through life cycle analysis.

Interdisciplinary courses at Suez Canal University, which emphasize environmental science, and sustainability, instruct students on green strategies such as the chemistry of recycling processes, renewable, safety, green energy, and photocatalysis, a process that takes advantage of Egypt's abundant sunlight [44]. Another goal of Suez Canal University environmental programs is to extend current GCE programs to the elementary and secondary school curriculums.

3.3.3.1 GCE-Egypt (SATLC)

Another important success in reforming Egypt's chemical education programs was undertaken between 2002 and 2011 by A. F. M. Fahmy, and J. J. Lagowski, who developed SATLAC, The Systemic Approach to Teaching and Learning [45–47]. This chemical education program addressed green organic chemistry education, with the goal of not only helping students better understand content but also to prepare them to meet global challenges. The program consisted of four courses, the first three of which dealt with aliphatic, aromatic, and heterocyclic chemistry, while the fourth was titled: GC in lab experiments for Faculty of Science Students. SATLAC was research-based, field-tested, used constructivist theory, and the model of multiple intelligences. Having demonstrated that students performed better using SATLAC than with conventional instruction, several Egyptian universities adopted the program. SATLAC was then disseminated at numerous conferences, seminars, and workshops, for example in: Istanbul, Turkey; Karachi, Pakistan; Mauritius, Algeria; Libya, and Syria. Two Jordanian-Egyptian Conferences were run during 2005–2006. In addition to SATLAC, a general chemistry course was also prepared for pre-university students. SATLAC was supported by the Italian Interuniversity Consortium, Chemistry for the Environment (INCA), part of the European TEMPUS educational program.

One important Egyptian green chemical educator is Dr. Salwa Elmeligie (Type II) who works at the intersection of pharmacy and chemistry at Cairo University. At the Pharma Middle East 2015 Conference, held November 2–4 in Dubai, UAE, Dr. Elmeligie presented a talk titled: "Greening the Pharmaceutical Industry to Afford Good Laboratory Practice." [48] In this seminar, she presented a roadmap for reforming the Egyptian and ME pharmacy education curriculum to include more sustainability and GC. Some of the practical goals that she advanced for the sake of better serving students were these:

1. Interactive teaching methods should be used in the classroom.
2. Instructors should introduce students to the latest GC developments.
3. New GC experiments should be developed.
4. Students should be encouraged to design green experiments and processes, and use green reagents.
5. Economize on lab water usage by for example, substituting vacuum for water aspirators.
6. Reduce solvents; this will have a domino effect in reducing waste, and generating cost savings.

In addition, she presented several strategies to integrate GC into an already crowded chemical curriculum, as for example by promoting the concept of GC to a wider community of scholars and students, using: awards, educational activities legislation, R&D grants, and fostering good relations with EPA-type organizations in each country.

Another important pedagogy that Elmeligie suggested was that instructors should present GC through the lens of environmental history (perhaps through case studies) to engage students and teach them how to learn from the global collective experience. Moreover, in the area of good laboratory practice, she explained how well-thought-out lab design of labs can streamline workflow, and help students work more efficiently and productively, while cutting waste.

In another talk at the 2015 Pharma Middle East Conference, Professor Elmeligie [49] extended the concept of GC to pharmaceutical chemistry, and thus created pedagogy for green pharmaceutical chemistry education. She advocated that GC principles not only be taught as a best practice in the field of pharmacy, but that students be taught that new drugs and other products be invented and manufactured using green chemical R&D techniques. To illustrate her ideas, she used the green synthesis of ibuprofen. Her talk, titled: "Green chemistry as a recent trend in Pharmacy education to afford Pharmaceutical products," was educational in nature, introducing the audience to reasons that support GC. However, she also indicated that GC contributes to employment, and to better products that are more affordable for all.

Numerous other Egyptian scientists are publishing GC research in scientific journals. For example, Suez Canal faculty member Ahmed Shahat has expertise in green inorganic chemistry and materials science chemistry, and according to ResearchGate [50], has authored 40 publications.

Cairo University is also a member of the PhosAgro/UNESCO/IUPAC Partnership [51] in the GC for Life Program, initiated in 2013, that challenges young scientists to perform breakthrough research for sustainable technologies in biochemistry, and interdisciplinary fields such as geochemistry, ecology, biotechnology, and healthcare.

3.3.4 Malta

In this section, GCE highlights will be taken from the Malta Conferences [52]. In 2003, the University of Malta introduced a group of first year students to GC in a GCE project run under the supervision of the Department of Educational Studies of the University of York (UK). Students participated in GC extra-curricular activities spread over a whole year. Students participated in seminars, teamwork, poster sessions, and small group discussions. Malta Conferences have been run every two years since, at various locations.

In a program [53] called: "Frontiers of Chemical Sciences: Research and Education in the Middle East," at the 7th Malta Conference, a visionary roadmap for research and education in the Middle East was established by bringing together a total of 15 Middle Eastern nations including Bahrain, Egypt, Iran, Iraq, Israel, Jordan, Kuwait, Lebanon, Libya, the Palestinian Authority, Qatar, Saudi Arabia, Syria,

Turkey, and United Arab Emirates. Participants studied the need for green and sustainable chemistry to overcome waste, pollution through alternative energy, and discussed the need of education for the purpose of attracting students to the field of chemistry. Moreover, the roadmap suggested reforming ME chemical education by making it greener, using exchange programs, forming a ME virtual campus, and extending sustainability to other fields such as pharmacy, toxicology, and clinical chemistry. Among the sponsors of Malta VII were UNESCO, ACS (American Chemical Society), the ACS Division of Chemical Education, AAAS (the American Association for the Advancement of Science), and the Committee of Concerned Scientists. Significant financial support was also received from the Carnegie Foundation of New York, the Rockefeller Brothers Fund, and the Alexander von Humboldt Foundation (Germany).

One example of GCE at Malta VII, was Rachel Mamlok-Naaman's (Weizmann Institute, Israel) seminar: *Learning About Sustainable Development in Socio-Scientific Issues-Based Chemistry Lessons on Bio-Plastics* [54]. In another presentation, Walter Kohn's [55] presentation on "A World Predominantly Powered by Solar and Wind Energy" indicated the enormous and urgent problems facing the planet Earth if alternative energy sources are not immediately adopted.

At Malta 3, Middle East participation was highlighted in IUPAC's Chemistry International publication [56]. Additionally, a workshop on Science Education and Green Chemistry was co-chaired by Boshra Awad (Egypt), Farouk Fahmy (Egypt), and Ann Nalley (USA), while the workshop on Alternative Energy Sources [57] was co-chaired by Hani Khouri (Jordan) and Hassan Zohoor (Iran).

3.3.5 Israel

In Israel, applied GC (Type I workers) garners more attention than academic GCE, but there are a few exceptions. For example, Shwartz et al. [58] studied the context and ramifications of GC in order to help high school students transfer and apply content knowledge to new situations. Both qualitative and quantitative critical reasoning skills were addressed. Results showed that through the project, students improved their cognitive skills. Another research question posed in the study was: Can a GC context-based curriculum increase higher-order cognitive skill learning? Students studied five online GC papers, and then completed five different tasks in two domains that involved GC: (a) chemistry content knowledge and (b) social-scientific. Students showed the most success in the later domain, by increasing their skills in: comparing, and in social-scientific argumentation, but students also increased their ability to communicate effectively with regard to scientific argumentation.

In another study, Hofstein et al. [59] described a curriculum that includes the primary goal of content learning while educating students to become scientifically literate and environmentally responsible citizens, by integrating sustainability into chemical education pedagogy. The author described a unique chemical education curriculum for preservice chemistry teachers that were designed to not only imbue the concept of science literacy and social responsibility, but to address issues of sustainability, the chemical industry, and the environment through case studies. While addressing environmental issues, students were exposed to different pedagogical strategies incorporating active learning through debates and field trips. Also, an internet website provided students with access to up-to-date information on industry, sustainability, and environmental issues. Recently, a module was developed and implemented in upper secondary classes, consisting of world-wide topics on water quality and global warming.

At Israeli Universities, all GC faculty are Type 1, mainly engaged in GC research. For example, in the school of chemistry at Tel Aviv University, Green Chemistry [60] is featured as an area of study, and three academic GC researchers are featured: Prof. Sergey Cheskis, Prof. Moshe Kol, and Prof. Arkadi Vigalok (who was a Member of the Scientific Advisory Board of the 1st Israel Conference on Green Chemistry, Tel Aviv 2005). At the Hebrew University of Jerusalem, both Li Malesis and Jochanan Blum perform organic GC research. On the other hand, at Bar Ilan University, GC is featured as part of the "Cleantech" curriculum which searches for sustainable substances.

The first GC Conference [61] held in Israel took place in 2007 at Tel Aviv University in its Porter School of Environmental Studies where the roles and future of both academic GC and industrial GC were discussed. The conference, entitled "Green Chemistry – Applications, Research and Trends," mainly addressed industrial GC with the goal of initiating more attention in academia and government work, to find breakthrough solutions to overcome pollution problems, including those caused by munitions producers, and energy generators.

Another way that GCE is indirectly being advanced in Israel is through the Israel Chemistry Society which annually awards its Green Chemistry Industry Prize [62]. Although the Prize is awarded to industry, in the award ceremonies, GCE is indirectly addressed, by for example, talks presented by leaders like John Warner. GCE is also given publicity through news media outlets.

3.3.6 Iran

Iran is one of the foremost GCE leaders in the ME, already claiming a long and famous scientific tradition that traces its roots to ancient Persia. With respect to GC, at the prestigious Iran University of Science and Technology, GC is a focal point of research where these professors are green chemistry experts [63]: Seyed Hashemianzadeh (green analytical chemistry), Mohammed Dekamin, Shahzad Javanshir and Mohammd Naimi-Jamal (green organic applications). Furthermore, the young scientist Dr. Mehdi Mohammadi is one of only six researchers who was awarded support from UNESCO's PhosAgro Green Chemistry for Life project [64], also supported by

IUPAC. It promotes breakthrough sustainable use and design of chemicals and chemical processes. Selected from among 119 applications, Dr. Mohammadi's proposal addressed the enzymatic production of biodiesel from waste oil by using two lipases covalently immobilized on a magnetic silica nanocomposite.

3.3.6.1 Iran GCE

In 2012, Bodlalo et al. [65], on the faculty of the Teacher Training University in Tehran, published an enlightening paper titled "Comparative Study in Green Chemistry Education Curriculum in America and China" that compares secondary GCE in Iran to that in the U.S. and China. Although in the U.S., GC is absent in the NGSS curriculum, nonprofit organizations like Beyond Benign [66] have taken up the cause, providing outreach, and has even worked to set up elective GC courses in high schools. However, in Iran, GC is already a compulsory subject in Grade 9 of Iran's innovative science curriculum that is designed to generate informed and scientifically literate citizens who both appreciate and understand that natural resources are finite, and that the environment should be free of waste and pollution. One goal is to educate students who as future citizens will be able to make sound judgments about the value of the environment versus industrialization. Iran's curriculum emphasizes the concept that GC is needed to overcome water and air pollution, to erase the environmental harm arising from pollution, and to stimulate recycling, and conserve energy. However, Bodlalo offers some disadvantages to Iran's GC curriculum compared to the U.S. One problem is that in Iran, laboratory work is deemphasized. Another issue is that in Iran, only a few GC principles are taught instead of all 12 like in the U.S. (although it could be argued that teaching all 12 GC principles would result in a cognitive overload for young secondary students, and so it is better to start with just a few GC principles). A third disadvantage to Iran's GC pedagogy is that, even though active learning teaching methods are espoused, in actual classroom practice, expository learning is mainly deployed. Fourthly, in Iran, assessment is based on exams and classroom activities that are not linked to national standards. Last of all, in Iran, assessments do not include a laboratory practical exam.

3.3.7 Saudi Arabia

Saudi Arabia is one of the world's largest producers of oil, while Iraq, Iran, United Arab Emirates (UAE) and Kuwait are also in the top ten. SABIC, the Saudi Arabia Basic Industries Corporation [67] is based in Riyadh, Saudi Arabia; it was the third largest chemical company as of 2014. The company produces industrial polymers such as: polyethylene, polyolefinics, and polypropylene, and chemicals such as ethylene glycol, MTBE, fertilizers, and metals. The petroleum and petrol-chemicals industries of the Gulf States have given rise to a host of environmental problems, so there is no question that GCE is needed. But the conundrum facing one of the world's top petroleum producers is that invoking carbon constraints on the Kingdom would trigger dire socioeconomic consequences, and some Saudis argue that social-economic factors should therefore be factored in when determining measures to combat climate change. Furthermore, the Saudi Arabian oil minister has indicated that the country has already vigorously implemented carbon-reducing measures [68], for example, by injecting carbon dioxide into the ground, using CO_2 as a polymer feedstock, and reducing overall carbon emissions from fossil fuel combustion sources, like automobiles. In addition, the Saudis have enacted some very strong environmental legislation.

Saudi Arabia currently imports a lot of its technical workers, but according to Nature magazine [69], the chemical research future is bright for Saudi Arabia with the advent of King Abdullah University of Science and Technology (KAUST) founded only in 2009. At KAUST, graduate-level chemical research is on par with that performed at premiere Western universities. Not mentioned in the article is the new joint venture with Dow Chemical, a company heavily invested in GC. How quickly Saudi Arabia transforms to a knowledge-based economy, will in large part depend on the quality of its universities.

One strategy to get scientists and educators to collaborate on GC at the intersection of academia, industry, government and education is to focus their efforts toward a designing and deploying GC, GCE, and sustainable strategies that protect both the environment and the health of the populations while also protecting economic sustainability.

3.3.7.1 GCE in Saudi Arabia

In a 2016, Ismail [70] reported a study on GC pedagogy, designed to instruct Saudi students about the importance of GC in industrial research as a way to build a sustainable society. In the study, the principles of GC were adopted in organic chemistry lab experiments for undergraduate students. The study group consisted of 100 students divided into four groups of 25 students each. By comparing the new greener labs to their non-green predecessors, the author found that GC imparted these advantages: a) students found the experiments more practical and simpler to execute, b) safety hazards were reduced by using microscale techniques, c) less waste was generated, and d) atom efficiency improved. Moreover, students were given pre- and post-tests on GC and sustainability, and results showed that students increased their general knowledge and understanding, and better appreciated the environmental ethics associated with GC. Discussion groups allowed students to share and reinforce GC knowledge. One of the study's objectives implemented was to "green" the existing traditional lab using GC principles. Most of the labs involved using reagents to identify compounds such as carboxylic acids and bases through qualitative functional groups procedures. One improvement was, for example, on the lab for the identification of aldehydes/ketones using the Schiff test; the amount of liquid volume used was reduced from 2 mL to 20 μL, thereby reducing reagent cost, providing a safer experience for students, and making it easier to for students to manipulate glassware and reagents while generating less waste.

Another example of GCE in Saudi Arabia was presented at the 12th Asian Chemical Congress in 2007 where Ali et al. [71] reported on the development and publishing of a lab manual devoted to green organic microscale experiments, a collaboration between chemistry and education faculty at three different Saudi universities. Development of the lab manual was supported by UNESCO, and it was published in many different languages, and also made available online. The lab manual was organized into two parts, on: (1) techniques in organic chemistry and (2) identification of organic functional groups. Part 1 addressed filtration, separation, melting/boiling points, distillation, crystallization, reflux, chromatography, separation of a mixture of organic compounds, the Lassaignes test, and modeling organic compounds. Part 2 covered: alkanes, alkenes, alkynes, halogens, alcohols, esters, carbonyls, carboxylic acids and their derivatives, and report writing. The original experiments were modified according to green chemical precepts, and pretested. Additionally, special workshops trained instructors on how to perform the experiments according to best practice.

At King Abdulaziz University (KAU), at the Sciences Faculty for Girls, green sustainable chemistry is researched by nine faculty members who specialize in organic chemistry, so that graduates can work in the large petrochemicals and chemicals industries [72]. KAU also has a community outreach program and runs a green video contest for kids [73].

Green research is being performed at other Saudi universities as well. For example, at King Saud University, GC protocols are developing environmentally benign solvents, and organometallic catalysts. For example [74], Abdulrahman Al-Warthan utilizes GC to synthesize nanomaterials, like, graphene oxide using plant extract. Medhat A. Shaker studies the removal of pharmaceutical contaminants from wastewater using nanomaterials. At the University of Jeddah [75], DeiaAbd E-Hady, performs green analytical research involving a novel separation of acrylamides by ionic liquid collapse capillary electrophoresis (ILCCE).

3.3.8 UAE

At the Petroleum Institute based in Abu Dhabi, capital of the United Arab Emirates, chemical engineering faculty [76] have learned how to integrate the concepts of GC and green chemical engineering into the curriculum. While their work isn't strict GCE, because it is not being done in a chemistry academic setting, it is quasi-GCE because training in GC is involved, and students are imbued with GCE concepts. All faculty (Type I GC) are listed in the Department of Chemical Engineering: Dr. Maaike Kroonin researches green solvents, while Dr. Samuel Stephen studies GC smart adsorbents. Moreover, the Department of Chemical Engineering offers several courses [77] touching on GC and sustainability. For example, the department offers CHEM 566, Construction Materials and Green Chemicals as a graduate level course in their curriculum; it surveys green chemicals that can be used in oil well drilling. CHEM 565 on fuels and alternative energy covers sustainable energy, and the disadvantages of fossil fuels, including carbon capture and sequestration of carbon dioxide. Another course, CHEM 560: ENVIRONMENTAL SCIENCE AND WATER TECHNOLOGY investigates environmental pollution caused by chemicals, and describes prevention methods.

Moreover, at the University of Sharjah [78], a Emirati private national university located in University City, Sharjah, United Arab Emirates, Dr. Kamrul Hasan investigates both renewable energy and GC catalysis as a organometallic chemist.

3.3.9 Bahrain

One exceptional ME GC educator is Dr. Saeed Al-Alwai who at the University of Bahrain, taught a GC course [79], and later organized workshop courses to instruct chemical workers on how to minimize waste, and utilize cleaner production practices, for example, at the Gulf Petrochemicals and Chemicals Association (GPCA) conferences.

3.3.10 Turkey

Several important international conferences incorporating GCE have been held in Turkey. One example is the 18th International Conference on Chemical Education that was held in Istanbul, Turkey in 2005 [80]. It featured GC, environment-friendly chemistry experiments, microscale chemistry, all contributed from experts across the ME and elsewhere.

At the 1st International Conference on Green Chemistry and Sustainable Technologies (2015) held in Izmir, Turkey, T. Gunter et al. [81] presented an interesting GCE Turkish college study titled: “The effect of problem based learning (PBL) on students’ comprehension levels in the subject of green chemistry and sustainability.” Researchers investigated how students construct green chemical knowledge using a problem-based learning approach in a analytical laboratory experiment on the qualitative analysis of cations. The control group ($N = 31$) performed the lab experiment using a traditional approach while the experimental group ($N = 63$) used PBL pedagogy, which modified the lab experiment by making students: perform research on a relevant problem, use the scientific method to devise hypotheses, collect information, and work in cooperative groups. The effect of the PBL lab exercise on students’ comprehension levels of GC and sustainability was studied using an assessment titled: “Green Chemistry and Sustainability Comprehension Level Test” which

was composed of open-ended questions deployed as pre- and post- tests. Based on the content analysis results, the authors concluded that their PBL pedagogy:
1. Increased student comprehension of GC and sustainability.
2. Increased student understanding and reduced misconceptions about GC and sustainability.
3. Could be applied to increase student understanding of GC and sustainability issues.
4. Could be applied to other situations to increase students' ecological and environmental awareness.

3.3.11 Palestine

In a poignant article [82] expressing the desire of millions living in the middle east, Z. M. Lerman published "Chemistry and Chemical Education as a Bridge to Peace." In this paper, the author advocated using chemistry education, including GCE, as a focal point on which Middle East nations can collaborate to solve their numerous problems. Her paper was presented at a Malta Conference, and is available on ResearchGate. Lerman's paper was visionary for it depicted the plight of the ME in context, for example, as being an important supplier of nonrenewable petroleum over which wars have been fought. She further elaborated the ME context by describing problems such as water scarcity and pollution through fossil fuel combustion, and how collaboration to solve such problems can bring nations together to find common solutions. She used many brilliant analogies to put across her point that the wider and more important issues of decreasing poverty are the ones that humanity should focus on. Her work has led to scientific collaborations between Palestinian and Israeli universities, and there have been numerous ongoing collaborations between scientists in Palestine, Israel, Kuwait, Iran, Jordan, and Egypt, especially in producing a database for water purification [83]. Moreover, Nobel Laureate Professor Roald Hoffmann held workshops for Middle Eastern graduate students from Jordan and Egypt [84]. As a result of GCE discussions, one idea suggested with respect to GCE was that GC should be integrated into ME science curricula using SATLC, and as a result, a collaborative effort resulted with the Israel Institute of Technology.

3.3.12 ChemRAWN

In 1976, IUPAC established a standing committee called ChemRAWN (Chemistry Research Applied to World Needs). ChemRAWN [85] conferences and projects to advance the concept of sustainability using chemical technologies have taken place across the globe. To extend the concept of sustainability to GC, in 2001 IUPAC assembled a work group on Synthetic Pathways and Processes in Green Chemistry [86], and created a interdivisional sub-committee on GC. In regard to the ME, the ChemRAWN Committee has internationalized its participation, by for example, adding Nadia Kandle from Egypt [87]. In addition, at the ChemRAWN XIX Conference, papers were presented on green catalysis, herbal medicines, and water in the ME. In addition, Mustafa Sözbilir became chair of the IUPAC standing committee on chemical education in 2014 and, Professor Ehud Keinan of Israel became an elected member to the IUPAC Bureau [88].

3.3.13 ME outreach in GC

It is also notable that during the International Year of Chemistry, GC outreach was performed by Abdelkrom Cheriti in Algeria [89] and Kuppusamy Uthaman Kuwait [90]; outreach is an important mechanism to disseminate GC, and there must be many more examples that are not widely reported.

3.4 Summary

This paper has reviewed the history of GCE in the Middle East within the ME context. ME nations are making progress in GCE, often with the help of agencies such as UNESCO, IUPAC, and INCA. Iran has made the bold move of adding GC to its secondary curriculum. By incremental additions of GCE programs in individual countries, a critical mass may soon be reached to move GCE-ME into a new phase of growth in which formal academic programs are instituted.

References

[1] Tundo P, Mammino L. Green chemistry in Africa, green chemistry series number 5. Venice, Italy: INCA, 2002.

[2] Tundo P, de Rosi R, Romero R. Quimica verde in Latino America, green chemistry series number 11. Venice, Italy: INCA, 2004.

[3] Lunin V, Tundo P, Lokteva E. Green chemistry in Russia, green chemistry series number 12. Venice, Italy: INCA, 2005.

[4] Canadian University of Dubai. Dec 2016 Available at http://www.cud.ac.ae/personnel/dr-hoshiar-nooraddin. Accessed:16 Dec.

[5] Idowu OS, Kasum SB, Mermod SY. People, planet, and profit. New York, NY: Rutledge, 2016.

[6] List of Middle Eastern countries by population. Dec 2016 Available at. Accessed:16 Dec https://en.wikipedia.org/wiki/List_of_Middle_Eastern_countries_by_population.

[7] World population projected to reach 9.7 billion by 2050. Dec 2016 Available athttp://www.un.org/en/development/desa/news/population/2015-report.html. Accessed:16 Dec.

[8] UN and Climate Change. Dec 2016 Available at. Accessed:16 Dec http://www.un.org/climatechange/blog/2015/03/will-weather-like-2050/.

[9] 10 projections for the global population by 2050. Dec 2016 Available athttp://www.pewresearch.org/fact-tank/2014/02/03/10. Accessed:16 Dec.

[10] Population Trends and Challenges in the Middle East and North Africa. Dec 2016 Available at http://www.prb.org/Publications/Reports/2001/PopulationTrendsandChallengesintheMiddleEastandNorthAfrica.aspx. Accessed:16 Dec.

[11] The World's Worst Pollution Problems. Dec 2016 Available at http://www.worstpolluted.org/files/FileUpload/files/WWPP_2012.pdf. Accessed:16 Dec.
[12] Simpson IJ, Aburizaiza OS, Siddique A, Barletta B, Blake NJ, Gartmner A, et al. Air quality in mecca and surrounding holy places in Saudi Arabia during Hajj: initial survey. Environ Sci Technol. 2014;48(15):8529–8537.
[13] Solid Waste management in Iraq. Dec 2016 Available at http://www.ecomena.org/tag/solid-waste-management-in-iraq/. Accessed:16 Dec.
[14] National solid waste management plan for Iraq. Dec 2016 Available at https://www.ncbi.nlm.nih.gov/pubmed/19470543. Accessed:16 Dec.
[15] Water scarcity in the Middle East and North Africa. Dec 2016 Available at http://hdr.undp.org/en/content/water-scarcity-challenges-middle-east-and-north-africa-mena. Accessed:16 Dec.
[16] Water management in Africa and the Middle East. Dec 2016 Available at http://collections.infocollections.org/ukedu/en/d/Jid13we/5.1.html. Accessed:16 Dec.
[17] Water rationing strategy to combat shortages. Dec 2016 Available at http://www.irinnews.org/feature/2008/03/24/water-rationing-strategy-combat-shortages. Accessed:16 Dec.
[18] Will Ethiopia's Grand Renaissance Dam dry the Nile in Egypt? Dec 2016 Available at http://www.bbc.com/news/world-africa-26679225. Accessed:16 Dec.
[19] Environmental Peace Building in the Middle East. Dec 2016 Available at http://blogs.ei.columbia.edu/2016/06/24/environmental-peace-building-in-the-middle-east/. Accessed:16 Dec.
[20] The Hydrogeology of Al Hassa Springs. Dec 2016 Available at https://www.researchgate.net/publication/267386708_The_Hydrogeology_of_Al_Hassa_Springs. Accessed:16 Dec.
[21] Highly toxic pesticide used in Ahsa date farms: Study. Dec 2016 Available at http://www.arabnews.com/node/383176. Accessed:16 Dec.
[22] Israel teachers the world how to grow food more wisely. Dec 2016 Available at http://www.israel21c.org/teaching-the-world-how-to-grow-food-more-wisely/. Accessed:16 Dec.
[23] Watergen. Innovative technology for water supply and air drying solutions. Dec 2016 Available at http://water-gen.com/. Accessed:16 Dec.
[24] Environmental problmes and perspectives in Middle East and North Africa Regions. Dec 2016 Available at http://www.iamb.it/share/img_new_medit_articoli/780_25maaroufi.pdf. Accessed:16 Dec.
[25] United Nations University. UNU-WIDER. Dec 2016 Available at https://www.wider.unu.edu/project/wiid-world-income-inequality-database. Accessed:16 Dec.
[26] Executive Summary; Green Chemistry. Pike Research. Dec 2016 Available at http://www.navigantresearch.com/wp-content/uploads/2011/06/GCHEM-11-Executive-Summary.pdf. Accessed:16 Dec.
[27] Green Chemicals will Save Industry 65 billion by 2020. Dec 2016 Available at http://green-chemistry.conferenceseries.com/. Accessed:16 Dec.
[28] UN Decade of Education for Sustianble Development 2005-2014. Dec 2016 Available at http://unesdoc.unesco.org/images/0014/001416/141629e.pdf. Accessed:16 Dec.
[29] United Nations Development Program. Human Development Reports. Education Index. Dec 2016 Available at http://hdr.undp.org/en/content/education-index. Accessed:16 Dec.
[30] Education Index. Dec 2016 Available at https://en.wikipedia.org/wiki/Education_Index. Accessed:16 Dec.
[31] UNESCO. UNITWIn UNESCO Chairs Programme. Dec 2016 Available at http://en.unesco.org/unitwin-unesco-chairs-programme. Accessed:16 Dec.
[32] UN The International Green Network. Dec 2016 Available at http://en.unesco.org/unitwin-unesco-chairs-programme. Accessed:16 Dec.
[33] Interuniversity Consortium. Chemistry and the Environment. Dec 2016 Available at http://www.incaweb.org/. Accessed:16 Dec.

[34] NATO-ASI Summer School on Green Chemistry. Dec 2016 Available at http://www.incaweb.org/education/nato-asi/. Accessed:16 Dec.
[35] NATO-ASI on Green Chemistry. Dec 2016 Available at http://www.incaweb.org/education/nato-asi/index.php. Accessed:16 Dec.
[36] New Organic Chemistry Reactions and Methodologies for Green Production. Dec 2016 Available at http://www.incaweb.org/education/nato-asi/index.php. Accessed:16 Dec.
[37] Summer School on Green Chemistry 10th Event. Dec 2016 Available at https://www.iupac.org/publications/ci/2008/3004/ca3_121008.html. Accessed:16 Dec.
[38] Summer School on Green Chemistry 1999. Dec 2016 Available at http://www.incaweb.org/education/summer_school_on_green_chemistry/pastss_1999.php. Accessed:16 Dec.
[39] UNITIN-MEGREC. Dec 2016 Available at http://virgo.unive.it/megrec/. Accessed:16 Dec.
[40] Greening Chemistry, The Mediterranean Green Chemistry Network. Dec 2016 Available at http://www.slideshare.net/UNESCOVENICE/greening-chemistry-megrec-objectives-and-activities-pietro-tundo-ca-foscari-university-of-venice-italy. Accessed:16 Dec.
[41] TEMPUS Programme. Dec 2016 Available athttp://eacea.ec.europa.eu/tempus/programme/about_tempus_en.php. Accessed:16 Dec.
[42] The Case for Green Chemistry. Integrating Sustainability into Curricula Campus. Dec 2016 Available at http://www.slideshare.net/davidhronek/scup-42-finalrev. Accessed:16 Dec.
[43] Ahmed MT. A new challenge for developing countries. Environ Sci Technol. 2005;12(2):114.
[44] Suez Canal University. New and Renewable Energy Authority. Dec 2016 Available at http://webcache.googleusercontent.com/search?q=cache:http://scuegypt.edu.eg/pages_media/2nd_Seminar_Agenda.pdf. Accessed:16 Dec.
[45] Fahmy AFM, Lagowski JJ. Systematic reform in chemical education: an international perspective. J Chem Educ. 2003;8(9):211.
[46] Lagowski JJ. Systemic approach to teaching and learning. J Chem Educ. 2005;82(2):1078–1083.
[47] Systematic Approach to Teaching and Learning. SATLAC in Egypt. Dec 2016 Available at http://old.iupac.org/publications/cei/vol3/0301x0an1.html. Accessed:16 Dec.
[48] Elmeiligie S. Greening the pharmaceutical industry to afford good laboratory practice. Dec 2016 Available at http://www.omicsonline.org/speaker/salwa-elmeligie-cairo-university-egypt/. Accessed:16 Dec.
[49] Elmeiligie S. Green chemistry as a recent trend in pharmacy education to afford pharmacy products. Dec 2016 Available at http://middleeast.pharmaceuticalconferences.com/speaker/2015/salwa-elmeligie-cairo-university-egypt-418565781. Accessed:16 Dec.
[50] Shahat A. Dec 2016 Available at https://www.researchgate.net/profile/Ahmed_Shahat3. Accessed:16 Dec.
[51] PhosAgro/UNESCO/IUPAC Partnership. Dec 2016 Available at http://www.unesco.org/new/en/natural-sciences/science-technology/basic-sciences/chemistry/green-chemistry-for-life/. Accessed:16 Dec.
[52] Malta Conferences Foundation. Dec 2016 Available at http://www.maltaconferencesfoundation.org/. Accessed:16 Dec.
[53] Frontiers in Science. Research and Education in the Middle East. Dec 2016 Available at http://www.divched.org/spring2016/Malta. Accessed:16 Dec.
[54] Division of Chemical Education. Dec 2016 Available at http://www.divched.org/spring2016/Malta. Accessed:16 Dec.
[55] A World Powered Predominantly by Wind and Solar Energy. Dec 2016 Available at http://aachen2050.isl.rwth-aachen.de/w/images/3/34/PIK_Global_Sutainability_22-27.pdf. Accessed:16 Dec.
[56] Malta III. Research and Education in the Middle East. Dec 2016 Available at https://www.iupac.org/publications/ci/2008/3003/cc4_081207.html. Accessed:17 Dec.

[57] Iran's Renewable Energy Potential. Dec 2016 Available at http://www.mei.edu/content/article/iran%E2%80%99s-renewable-energy-potential. Accessed:17 Dec.
[58] Schwartz Y, Ben-Zvi R, Hofst A. The use of scientific literacy taxonomy for assessing the development of chemical literacy among high-school students chemistry. Educ Res Pract. 2006;7(4):203–225.
[59] Hofstein A, Shwartz Y, Kesner M. Incorporation of sustainability into chemistry education: A 25-year-long partnership between chemistry education and Israeli industry. Abstracts of Papers, 247th ACS National Meeting & Exposition, Dallas, TX, United States; 2014. 16–20Mar2014 2014.
[60] Green Chemistry. Dec 2016 Available at. Accessed:17 Dec https://en-exact-sciences.tau.ac.il/chemistry/green_chemistry.
[61] Green Chemistry Promises a Cleaner Country. Dec 2016 Available at http://www.jpost.com/Business/Business-Features/Green-Chemistry-promises-a-cleaner-country. Accessed:17 Dec.
[62] The Israeli Chemical Society. Dec 2016 Available at. Accessed:17 Dec http://www.chemistry.org.il/ics-prizes.
[63] Iran University of Science and Technology School of Chemistry. Dec 2016 Available at http://www.iust.ac.ir/index.php?sid=20&slc_lang=en. Accessed:17 Dec.
[64] Six young scientists to receive Green Chemistry for Life Grants in 2015. Dec 2016 Available at http://en.unesco.org/news/6-young-scientists-receive-green-chemistry-life-research-grants-2015. Accessed:17 Dec.
[65] Bodlalo LH, Sabbaghan M, Jome SMRE. A comparative study in green chemistry education curriculum in America and China, 6th university conference on teaching and learning. Procedia Soc Behav Sci. 2013 October 10;90:288–292 10 October.
[66] Beyondbenign. Green Chemistry Education. Dec 2016 Available at. Accessed:17 Dec http://www.beyondbenign.org/.
[67] SABIC. Chemistry that matters. Dec 2016 Available at. Accessed:17Dec https://www.sabic.com/corporate/en/.
[68] Saudi Arabia to ratify Paris climate deal before Marrakech talks. Dec 2016 Available at http://www.climatechangenews.com/2016/10/31/saudi-arabia-to-ratify-paris-climate-deal-before-marrakech-talks/. Accessed:17 Dec.
[69] A 21st century transformation. Dec 2016 Available at http://www.nature.com/nature/journal/v532/n7600_supp_ni/full/532S4a.html. Accessed:16 Dec.
[70] Ismail OMS. Green approach for chemical education in chemistry lab. Am J Chem. 2016;6(2):55–59.
[71] Green technology for microscience experiments. Dec 2016 Available at http://marz.kau.edu.sa/Files/0001866/Researches/22443_38882.pdf. Accessed:17 Dec.
[72] Science Faculty for Girls. Green Sustainable Chemistry. Dec 2016 Available at http://gce-sgcg.njb.kau.edu.sa/Default.aspx?Site_ID=363800&lng=EN. Accessed:16 Dec 2016.
[73] Walk on Green for Contest. Dec 2016 Available at http://cees.kau.edu.sa/Pages-WOG-contest4-En.aspx. Accessed:17 Dec.
[74] Al-Warthan A. Dec 2016 Available at http://king-saud.academia.edu/AbdulrahmanAlWarthan. Accessed:17 Dec.
[75] Past and present research systems of green chemistry. Dec 2016 Available at http://green-chemistry.conferenceseries.com/speaker/2015/deiaabd-el-hady-university-of-jeddah-saudi-arabia. Accessed:17 Dec.
[76] The Petroleum Institute. Dec 2016 Available at. Accessed:17Dec http://www.pi.ac.ae/index.php.
[77] Graduate courses. Dec 2016 Available at http://www.pi.ac.ae/PI_ACA/pgp/courses.php. Accessed:17 Dec.

[78] Hassan K.. Dec 2016 Available at http://www.sharjah.ac.ae/en/academics/Colleges/Sciences/dept/ch/Pages/ppl_detail.aspx?mcid=23. Accessed:17 Dec.

[79] Al-Alwai S.. Dec 2016 Available at http://gpcaresponsiblecare.com/?pt_speakers=dr-saeed-al-alawi. Accessed:17 Dec.

[80] Proceedings of the 18th International Conference on Chemical Education. Dec 2016 Available at http://old.iupac.org/publications/cei/vol6/. Accessed:17 Dec.

[81] 1st International Conference on Green Chemistry and Sustainable Technologies. Dec 2016 Available at http://iupab.org/archived-conferences/2015/1st-international-conference-on-green-chemistry-and-sustainable-technologies/. Accessed:17 Dec.

[82] Chemistry and chemical education as a bridge to peace. Dec 2016 Available at https://www.researchgate.net/publication/226317033_Chemistry_and_Chemical_Education_as_a_Bridge_to_Peace. Accessed:16 Dec.

[83] Malin JM. Frontiers of chemical sciences – research and education in the Middle East. Chem Int. 2004;26(3):7–9.

[84] Science for peace in the Middle East. Dec 2016 Available at http://pubs.acs.org/cen/science/83/8351sci2.html. Accessed:17 Dec.

[85] Chemistry Research Applied to World Needs. Dec 2016 Available at https://iupac.org/who-we-are/committees/committee-details/?body_code=021. Accessed:17 Dec 2016.

[86] Tundo P, Anastas P, Black D, Breen J, Collins T, Memoli S, et al. Synthetic pathways and processes in green chemistry. Introductory overview. Pure Appl Chem. 2000;72(7):1207–1228.

[87] Reports from San Juan. Dec 2016 Available at https://www.iupac.org/publications/ci/2012/3401/3_sanjuan.html. Accessed:17 Dec.

[88] President's column. Dec 2016 Available at https://www.degruyter.com/downloadpdf/j/ci.2016.38.issue-1/ci-2016-0103/ci-2016-0103.xml. Accessed:17 Dec.

[89] International Year of Chemistry. Activities in Algeria. Dec 2016 Available at https://www.iupac.org/publications/ci/2012/3401/3_sanjuan.html. Accessed:17 Dec.

[90] International Year of Chemistry. Activities in Kuwait. Dec 2016 Available at http://www.chemistry2011.org/participate/activities?country=Kuwait&view=country_list. Accessed:17 Dec.

Jonathan Stevens

4 Virtually going green: The role of quantum computational chemistry in reducing pollution and toxicity in chemistry

Abstract: Continuing advances in computational chemistry has permitted quantum mechanical calculation to assist in research in green chemistry and to contribute to the greening of chemical practice. Presented here are recent examples illustrating the contribution of computational quantum chemistry to green chemistry, including the possibility of using computation as a green alternative to experiments, but also illustrating contributions to greener catalysis and the search for greener solvents. Examples of applications of computation to ambitious projects for green synthetic chemistry using carbon dioxide are also presented.

Keywords: green, quantum chemistry, catalysis, solvents, computational chemistry

4.1 Introduction

While modern quantum mechanical models of electrons in atoms and molecules arose in the early twentieth century [1], molecular quantum calculation of molecules first began to become accessible to non-theorists with the first release of John Pople's Gaussian package in 1970 [2]. In the intervening period, advances in theory and in computing technology have made computational chemistry, specifically quantum mechanical calculations of electrons in molecules, into a widely applied technique for research in spectroscopy, thermodynamics, and the mechanism and kinetics of chemical reactions [3]. Modeling of systems within solution is now a common practice [4].

Green chemistry is research or other activity focused on the reduction of toxicity in the practice of chemistry. Unlike "environmental chemistry", an appellation which may be used to describe a wide range of research projects, green chemistry is generally recognized as being defined by a set of twelve guiding principles [5]. In its current state, computational chemistry plays a significant role in "green" research. This chapter does not provide an exhaustive list of all of the contributions of quantum chemical computations to green chemistry, but it will show a number of intriguing examples. Computational chemistry is integral to research efforts seeking greener catalysts and to finding greener methods of testing potential catalysts for efficacy. It plays a role in investigations toward greener solvents for reactions and extractions. It provides thermodynamic and kinetic information on novel efforts to

https://doi.org/10.1515/9783110445923-004

transform carbon dioxide emissions from a source of pollution to a renewable feedstock for production of useful polymers. These examples will be discussed in detail in the following sections.

4.2 Greening catalysis – enzyme design

The ability to study chemical reaction pathways computationally provides a means to test reaction catalysts *in silico* before moving on to test promising candidates in laboratory synthetic efforts. Research of this sort has appeared in recent years in eminent publications such as Science [6], and a very recent volume of Accounts of Chemical Research has been devoted to the topic of "Computational Catalysis for Organic Synthesis" [7]. The goal of such synthetic research may be to improve stereoselectivity of products or increases in yield [6], but the methodology often simultaneously provides a possible "greening" of the synthetic process. For example, a recent article in Science [6] concerns design of an enzyme catalyst for a reaction familiar to any student of organic chemistry, the Diels-Alder synthesis. This work furthers the goal [8] of using organic catalysts to catalyze Diels-Alder syntheses, as opposed to traditional catalysts, which include toxic Lewis Acids such as boron trifluoride or tin tetrachloride [9–11]. The solvent used for experimentally implementing this enzyme is water [6], and so this work provides a very "green" example of a Diels-Alder synthesis. The Diels-Alder reaction selected is the reaction of 4-carboxybenzyl trans-1,3-butadiene-1-carbamate as the diene and with N,N-dimethylacrylamide as the dienophile. The reaction is displayed in Figure 4.1.

The work is an implementation of the Rosetta methodology [12] in enzyme design, in which the transition state and active site for the reaction going from substrate to product is located by quantum mechanical molecular orbital

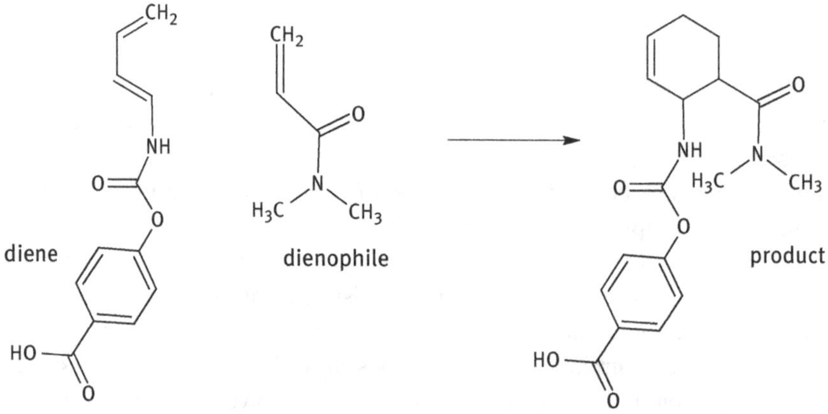

Figure 4.1: Diels-Alder reaction chosen for reference 6.

calculations and then structurally matched into a protein “scaffold” obtained from a database. The design in this work creates a “pocket” active site into which both substrates are held in an optimal encounter geometry for reaction and positioning of a hydrogen bond acceptor to interact with the NH moiety on the carbamate diene and a hydrogen bond donor to interact with the carbonyl on the dienophile. The hydrogen bond to the carbamate raises the energy of the homo on the diene, while the hydrogen bond to the dienophile stabilizes its LUMO; the net effect, in addition to holding the two molecules in position for reaction, is a reduction of the energy difference between the LUMO and the HOMO, resulting in a faster reaction rate. Siegal et al. [6] estimate that the hydrogen bonding lowers the energy of the transition state by approximately 4.7 kcal/mol.

Insertions into available protein scaffolds initially produced 84 prospective catalytic enzymes, two of which were experimentally found to have catalytic ability; additional improvements in catalytic activity were obtained by mutating the structures of enzymatic residues around the transition state. A feature of the design of the active site is that it is very stereospecific, this is in fact observed in the experimental distribution of products (>97 % for the specific isomer).

Enzyme design using a computationally designed active site is often referred to as a “theozyzme” approach [13, 14]; there is also a more empirical “minimalist “approach which does not employ quantum mechanical calculations and instead attempts to design *de novo* enzymes by matching desired function to known details of enzyme structures [15]. This latter approach tends to produce less efficient catalysts than the quantum-computational approach,’ but it is less computationally expensive; recent work discusses this trade-off, and illustrates the possible efficacy of the minimalist approach by presenting a retroaldolase catalyzing the reverse aldol condensation of methodol [16].

With respect to green chemistry, catalysis with engineered enzymes offers the possibility of replacing toxic catalysts with non-toxic protein catalysts; additionally, the solvent for enzymatic catalysis is water, clearly a greener choice than organic solvents [16].

4.3 Greening catalysis – “in silico” experiments

A very common computational approach to studying the kinetics and catalysis of a reaction is the use of quantum density functional methods such as B3LYP [17] to determine the structure and energies of the reactants, products and transition state of a reaction pathway; these calculations provide predictions of the thermodynamics of the reaction and of the energetics of the barrier to reaction arising at the transition state. This data may then be used to assess the kinetics of the reaction. Recent examples of this approach have been used to replace or inform experiments using metal catalysts such molybdenum [18], rhodium [19] or ruthenium [20]. Such work lends mechanistic insight and other information which ultimately may reduce the number of experiments conducted in catalytic research; from a green perspective, this means some reduction in the use, and need for disposal of, catalysts, reagents, and organic solvents.

A potentially more significant opportunity for "greening" catalysis research, however, appears in a recent work by Wheeler et al. [21] Recent research in the field of organocatalysis seeks to use computational chemistry to develop methods to screen catalyst designs for stereoselectivity, hence eliminating the need to synthesize and test many catalysts which would ultimately fail to provided desired degrees of stereoselection [21, 22, 25]. The work discussed here [21] is motivated by a desire to understand the effect of non-covalent interactions on Lewis-Base alkylation reactions, focusing in particular upon the allylation or propargylation of benzaldehyde by means of allyl silanes, which produces substituted alcohols as shown in Figure 4.2.

This reaction is known to be catalyzed by pyridine N-oxide and N,N′-dioxide catalysts, with varying degrees of enantioselectivity on the hydroxyl-substituted carbon. For several catalysts for these alkylation reactions, computational research was able to produce a predicted distribution of enantiomers which matched the experimentally observed product distribution. A challenging aspect of this work was that a particular catalyzed alkylation reaction might have 10 or even 20 relevant transition states connecting the reactant to the chiral products; these transition states are specific to formation of either R or S products, and predicting the distribution of enantiomeric products requires computations to determine all the structures and energies of all these transition states.

The task of predicting stereoselectivities for a large number of bipyridine N,N′-dioxide catalysts would then involve the computation of hundreds of transition states, a trying and time-consuming task. Sparks and Wheeler et al. were able to automate this work with the development of a computational tool kit referred to as AARON (Automated Alkylation Reaction Optimizer for N-oxides) [21, 26]. Testing of the AARON package revealed that in the case of allylations it was able to reproduce the experimentally known steroselectivities of a set of 18 bipyridine

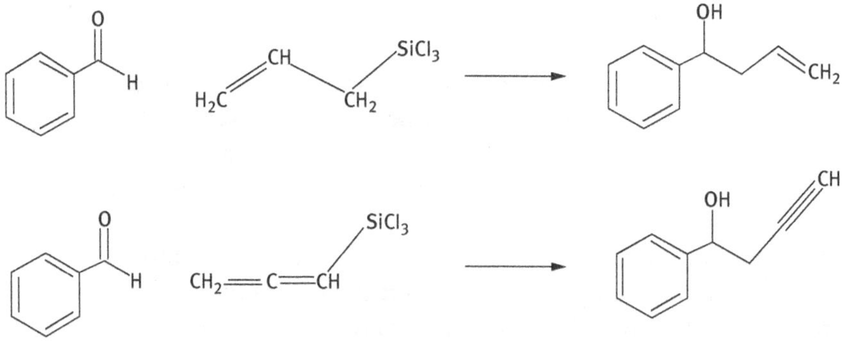

Figure 4.2: Allylation (top) or propargylation (bottom) of benzaldehyde discussed in reference 19.

N,N′ dioxide catalysts [21, 27]. The package has since been used to predict the results of propargylations with 100 new bipyridine N,N′ dioxide catalysts. Some of these have a high degree of predicted enantioselectivity; the authors hope there will be experimental testing of some of their predictions. Meanwhile, Sparks and Wheeler are developing a new version of AARON [21, 28] capable of dealing with a broad variety of organocatalyzed organic reactions. The implications for "greening" here are apparent; a future in which literally hundreds of experiments, with attendant use of materials and generation of chemical wastes, might be able to be replaced by implementation of a quantum computational package.

4.4 Toward greener solvents

Specific applications of quantum computational methods to green chemistry appear regularly in the Royal Society of Chemistry journal *Green Chemistry*. Much of this research is devoted to reducing usage of harmful or toxic solvents. An example of recent work [29] explores alternative solvents for the Mannich reaction [9, 29], the reaction of an amine, aldehyde, and enolizable ketone to produce a β-amino carbonyl compound, or Mannich base, as shown in Figure 4.3. This reaction plays an role in synthesis of bioactive species and pharmaceuticals [29, 30], but the synthesis is commonly run in organic solvents such as benzene, THF, acetone, and DMSO [29, 30].

Computational work [29] explores green catalysis of the Mannich reaction. Nornicotine, a natural product produced within tobacco plants, is a known catalyst for the aldol condensation in water [31], illustrating its ability to function as a catalyst in an aqueous medium. It is chosen for investigation as a green catalyst in aqueous solvent in this work. The computational work implements M062X hybrid density functional theory [32] and model solvents with the self-consistent reaction field (SCRF) in the form of the polarizable continuum model [33]. Investigation of the reaction potential energy surface finds that the energetic profile of intermediates and transition states along the reaction path with this catalyst are nearly identical in aqueous and DMSO or other organic solvent.

Another recent intensive use of quantum chemistry [34] has combined density functional calculations of the conformers of artemisinin (see Figure 4.4) with the COSMO_RS statistical mechanics method [35, 36] to a search for greener solvents for artemisinin extraction.

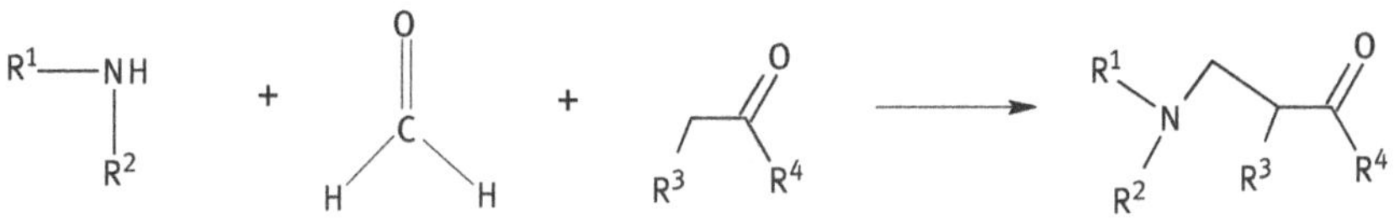

R^1—NH, R^2 + H–C(=O)–H + R^3–CH₂–C(=O)–R^4 → R^1R^2N–CH₂–CH(R^3)–C(=O)–R^4

Figure 4.3: The Mannich reaction.

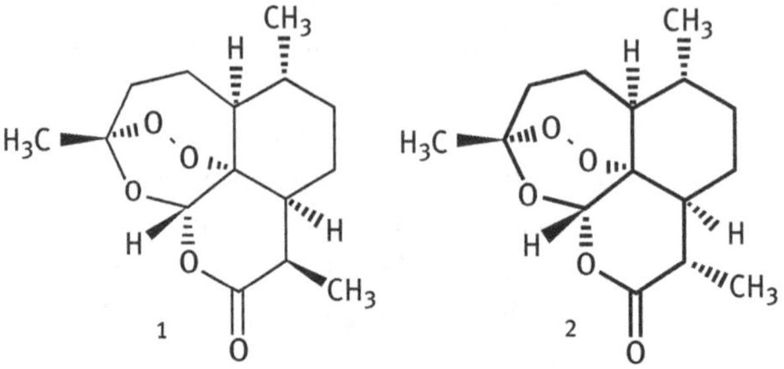

Figure 4.4: Structure of Artemisinin (1) and its diasteromer epiartemisinin (taken from reference 35).

Artemisinin is an antimalarial commonly extracted from the sweet wormwood plant [37] from which it is commonly extracted using hexane solvents [34, 38] or petroleum ether [34, 39], a mixture of hydrocarbons containing hexanes [40]. Newer approaches for extraction use fluorocarbons or the iodated fluorocarbon iodotrifluoromethane [34], compounds which have significant global warming potential [34, 41] as well as some environmental toxicity [34, 42].

The combined quantum mechanical/statistical mechanical modeling was employed to calculate the solubility of artemisinin, as well as its diastereomer, epiartemisinin, in a wide range of solvents. The predicted values for water and a number of common organic solvents were compared to available experimental values, finding generally good agreement in most cases, such as water, alcohols, and chloroform. Further calculations obtained predicted solubilities in a large number of "novel" solvents, for which little experimental data is available. Diacetone alcohol, DMPU (1,3-Dimethyl-3,4,5,6-tetrahydro-2(1H)-pyrimidinone), DMI (1,3-Dimethyl-2-imidazolidinone) and organic carbonates are predicted to be effective solvents with low toxicity. DMPU, glyceryl carbonate, ethyl lactate, and glycerol isobutyral in particular were identified as solvents that could be obtained from renewable feedstocks.

4.5 Polymer production from CO_2 emissions

As the quantity of atmospheric carbon dioxide increases, there is an increase in concern for the implications for global warming and climate change. As a result, recent quantum-computational investigations have been concerned with remediation of increasing concentrations of atmospheric carbon dioxide. While some computational studies deal with capture and storage of carbon dioxide [43], other investigations deal with the green approach of finding uses for CO_2 as a feedstock for syntheses of chemical products, a feedstock conceivably renewed by ongoing use of the otherwise polluting emissions from the combustion of fossil fuels. One example of such work may be found in a recent *Green Chemistry* publication involving the

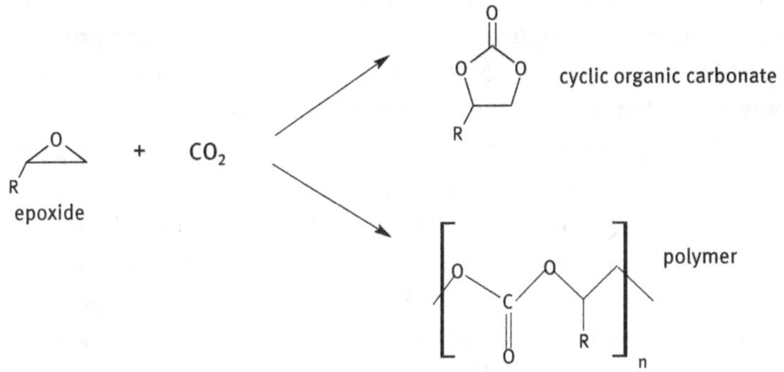

Figure 4.5: Reactions of CO_2 with epoxides.

carboxylation of organoboron compounds by means of catalysis with rhodium complexes; density functional theory is used to compare the relative efficacies of catalysis when diboron and diphosphane ligands are used within the rhodium complex [44].

An intriguing line of research in CO_2 utilization has focused on the reaction of carbon dioxide with epoxides to produce commercially useful polymers or cyclic organic carbonates; which, as we have seen, have uses as green solvents. The reactions are illustrated in Figure 4.5.

The polymer formation reaction has long been known to be catalyzed by complexes of zinc ions [45], but relatively more recent experimental work [45] has focused on the mechanism of the reaction when catalyzed by complexes of Cr^{3+}, Co^{3+}, or Al^{3+}, with bis(salicylaldimine) ligands; the reaction begins with complexation of the epoxide to the metal, followed by ring opening with an anion initiator such as Cl^- or N_3^-; this is followed by insertion of CO_2 and then alternating epoxide and CO_2 insertion as shown in Figure 4.6.

Research in computational chemistry has provided a significant amount of information about this process. Experimental data on the thermodynamics of

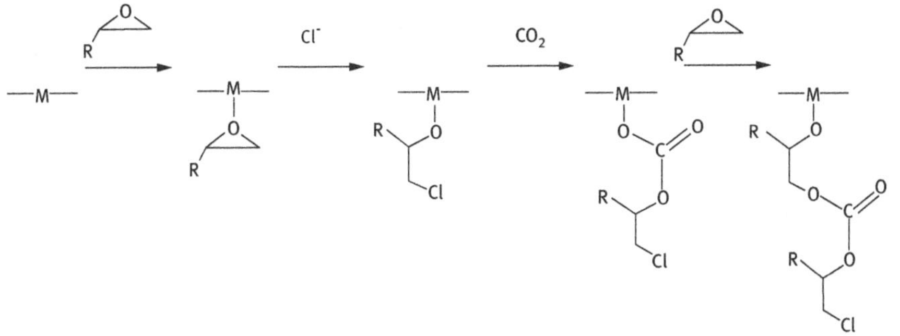

Figure 4.6: Catalyzed copolymerization of CO_2 and epoxide. Cl^- is shown as the initiator.

epoxide formation are sparse, but a quantum computational study has been made of the thermodynamics of CO_2-epoxide copolymerization, and of the competing process, carbonate formation, for a number of candidate epoxides [46]. In general, the formation of the polymer is found to be the more exothermic of the two reactions, but calculations of free energy changes for polymerization and carbonate formation indicate the formation of cyclic carbonates is generally more thermodynamically favored, due to entropy considerations; the authors note that this makes the formation of carbonate byproduct temperature sensitive. Consideration of computational work [46] with experimental data on metal-catalyzed reactions [47, 48] suggests that the competing carbonate formation most likely proceeds from a carbonate back biting reaction which proceeds most rapidly in a "metal-free" fashion as opposed to via a metal complex, as shown in Figure 4.7. The reaction is found to be in general exergonic for a number of epoxide substrates.

While carbonate back-biting is expected to be a degradation mechanism in ambient conditions, with CO_2 available, alkoxide back-biting reactions may cause depolymerization in more basic conditions [46, 49], these alkoxide reactions may cause cyclic carbonate formation (Figure 4.8) or complete decomposition to epoxides (Figure 4.9). In general, for a number of copolymers, calculations determine that barriers to the formation of epoxide are higher than those for the formation of cyclic carbonate.

Quantum computations determine that epoxide forming reactions are thermodynamically favorable; however, cyclic carbonate is in general the only product because epoxide formation has a higher barrier for reaction. In addition, the epoxide-forming reaction gives rise to carbonate polymers that undergo carbonate back-biting.

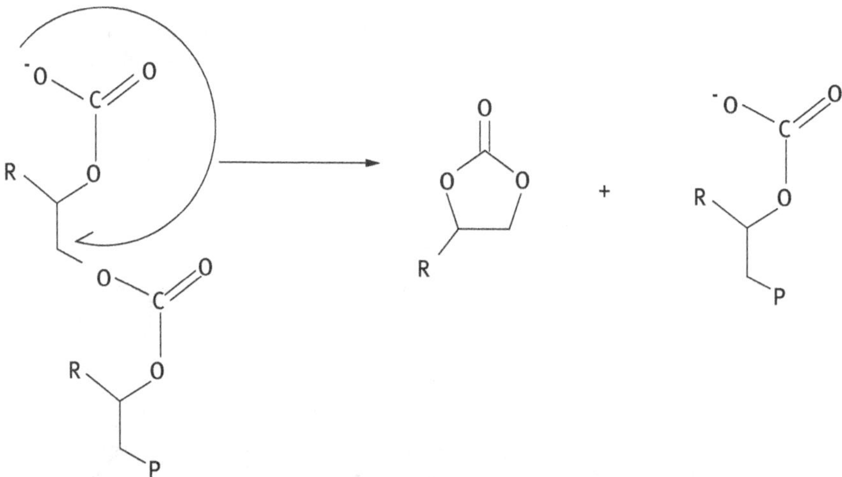

Figure 4.7: Carbonate back-biting reaction. "P" indicates polymer chain.

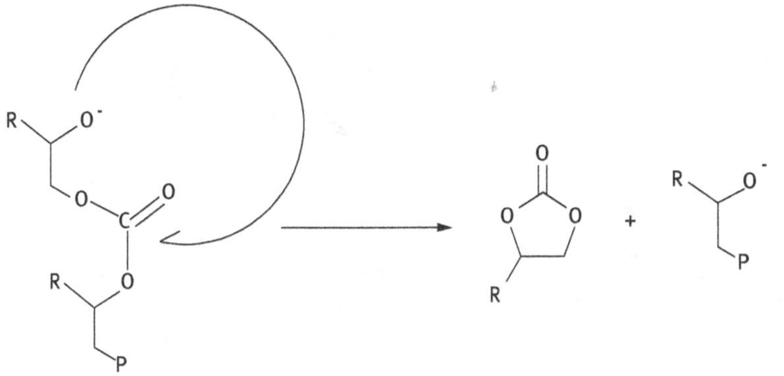

Figure 4.8: Alkoxide "back-biting" reaction producing a cyclic organic carbonate. "P" represents polymer chain.

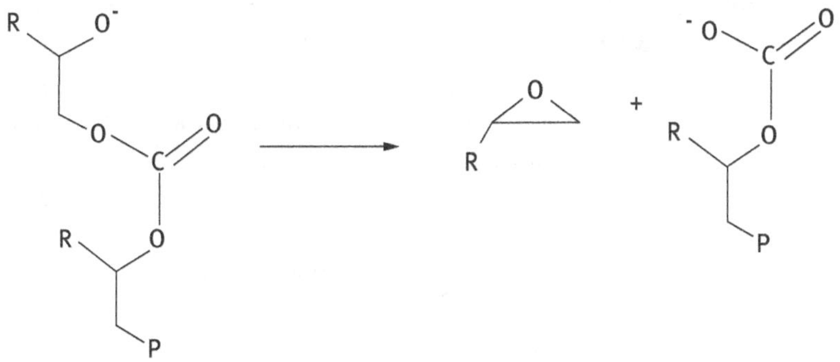

Figure 4.9: Alkoxide back-biting reaction producing an epoxide. "P' represents polymer chain.

Similar calculations [50] in conjunction with experiments have examined the impact of different epoxide substrates, in particular 1,4 and 1,3-cyclohexadiene oxides, or 1,2-epoxy-4-cyclohexene and 1,2-epoxy-3-cyclohexene. Computations compared 1,2-epoxy-3-cyclohexene, 1,2-epoxy-4-cyclohexene and cyclohexene oxide (1,2 epoxy cyclohexane). Experiments show that 1,2-epoxy-3-cyclohexene is the more reactive species than 1,2-epoxy-4-cyclohexene for formation of products, computation explains that this is the result of lower barriers to the epoxide ring-opening step involved in both polymerization and carbonate formation.

Experiments indicate that the reaction of 1,2-epoxy-3-cyclohexene produces copolymer and cis-cyclic carbonate, but no trans cyclic carbonate is formed (see Figure 4.10) The calculations show that 1,2-epoxy-3-cyclohexene is unusual in that polymer formation is thermodynamically more favorable than trans-cyclic carbonate formation; in particular *trans* carbonate formation is endergonic in the case of 1,2-epoxy-3-cyclohexene substrate.

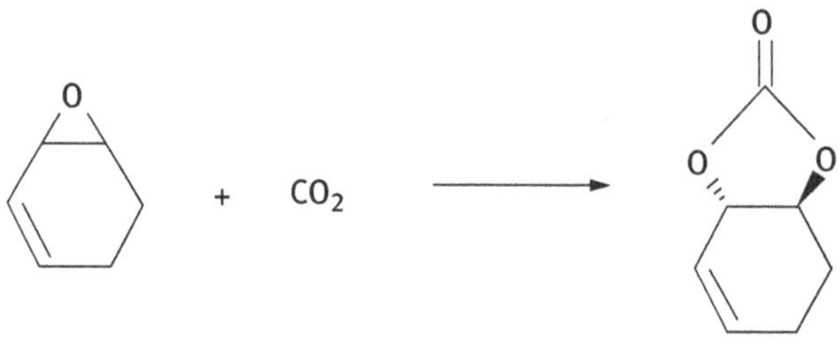

Figure 4.10: Hypothetical reaction of 1,2-epoxy-3-cyclohexene to produce a trans cyclic carbonate. This product is not observed experimentally, the corresponding *cis* product is.

4.6 Conclusion

Quantum computational chemistry has been shown to be an important contributor to green chemistry in several promising and exciting ongoing developments. In synthetic chemistry, the ability to study reactions and catalysis "*in silico*" increasingly offers a means of replacing some experimental laboratory research. Quantum computation has also been shown to offer information on choosing greener solvents. Computation plays a key role in advancing prospects for the design of enzyme catalysts; these catalysts offer the hope of replacing toxic catalysts and organic solvents with non-toxic organic catalysts and aqueous solvents. Molecular orbital quantum calculations play an ongoing role in green methods for the remediation and recycling of carbon dioxide.

As stated previously, this chapter does not attempt to discuss all applications of quantum calculations to making chemistry "greener". The journal *Green Chemistry* regularly provides accounts of such efforts, and the importance of green chemistry assures that other examples will continue to appear in other prominent journals.

References

[1] Born M, Oppenheimer RJ. "Zur Quantentheorie der Molekeln" [On the quantum theory of molecules]. Ann Der Phys. 1927;389:457–522.

[2] Hehre WJ, Lathan WA, Ditchfield R, Newton MD, Pople JA. Gaussian 70. Wallingford, CT: Gaussian, Inc.. Quantum Chemistry Program Exchange, Program No. 237 1970.

[3] Hehre WJ. A Guide to Molecular Mechanics and Quantum Chemical Calculations. Irvine, CA: Wavefunction, Inc., 2003. Available at: https://www.wavefun.com/support/AGuidetoMM.pdf. Accessed: 15 Dec 2016.

[4] Skyner RE, McDonagh JL, Groom CR, Van Mourik T, Mitchell JB. A review of methods for the calculation of solution free energies and the modelling of systems in solution. Phys Chem Chem Phys. 2015;17:6174–6191.

[5] Warner Babcock Institute for Green Chemistry. The 12 principles. Available at: http://www.warnerbabcock.com/green-chemistry/the-12-principles/. Accessed: 15 Dec 2016.

[6] Siegel JB, Zanghellini A, Lovick HM, Kiss G, Lambert AR, St. Clair JL, et al. Computational design of an enzyme catalyst for a stereoselective bimolecular Diels-Alder reaction. Science. 2010;329:309–319.

[7] Tantillo D. Computational catalysis for organic synthesis. Acc Chem Res. 2016;49:6.
[8] Ahrendt KA, Borths CJ, MacMillan DW. New strategies for organic catalysis: the first highly enantioselective organocatalytic Diels-Alder reaction. J. Am. Chem. Soc. 2000;122:4243–4244.
[9] Stretiweiser A, Heatchcock CH. Introduction to organic chemistry, 3rd ed. New York: MacMillan Publishing Company, 1985.
[10] Boron trifluoride, Center for Disease Control, The National Institute for Occupational Safety and Health (NIOSH). Available at:http://www.cdc.gov/niosh/idlh/7637072.html. Accessed: 15 Dec 2016.
[11] Tin Tetrachloride. National Institute of Health Pubchem., 2016 Dec 15 Accessed:15 Dec 2016. Available at:http://www.cdc.gov/niosh/idlh/7637072.html.
[12] Zanghellini A, Jaing L, Wollacot AM, Gong C, Meiler J, Althoff EA, et al. New algorithms and an in silico benchmark for computational enzyme design. Protein Sci. 2006;12(15):2785–2794.
[13] Tantillo DJ, Houk KN. Theozymes and compuzymes: theoretical models for biological catalysis. Curr Opin Chem Biol. 1998;2:743–750.
[14] Kries H, Blober R, Hilvert D. De novo enzymes by computational design. Curr Opin Chem Biol. 2013;17:1–8.
[15] DeGrado WF, Wasserman ZR, Lear JD. Protein design, a minimalist approach. Science. 1989;243:622–628.
[16] Raymond EA,, Mck KL, Yoon JH, Moroz OV, Moroz YS, Korendoych IV. Design of an allosterically regulated retroaldolase. Protein Sci. 2015;24:561–570.
[17] Becke AD. Density-functional thermochemistry. III. The role of exact exchange. J Chem Phys. 1993;98:5648–5652.
[18] Tanaka H, Nishibayashi Y, Yoshizawa K. Interplay between theory and experiment for ammonia synthesis catalyzed by transition metal complexes. Acc Chem Res. 2016;49(5):987–995.
[19] Park Y, Ahn S, Kang D, Baik M-H. Mechanism of Rh-catalyzed oxidative cyclizations: closed versus open shell pathways. Acc Chem Res. 2016;49(6):1263–127.
[20] Zhang X, Chung L-W, Wu Y-D. New mechanistic insights on the selectivity of transition-metal-catalyzed organic reactions: the role of computational chemistry. Acc Chem Res. 2016;49(6):1302–1310.
[21] Wheeler SE, Seguin TJ, Guan Y, Doney AC. Noncovalent interactions in organocatalysis and the prospect of computational catalyst design. Acc Chem Res. 2016;49:1061–1069.
[22] Chen J, Captain B, Takenaka N. Helical chiral 2,2′-bipyridine N-monoxides as catalysts in the enantioselective propargylation of aldehydes with allenyltrichlorosilane. Org Lett. 2011;13:1654–1657.
[23] Lu T, Zhu R, An Y, Wheeler SE. Origin of enantioselectivity in the propargylation of aromatic aldehydes catalyzed by helical N-oxides. J Am Chem Soc. 2012;134:3095–3102.
[24] Lu T, Porterfield MA, Wheeler SE. Explaining the disparate stereoselectivities of n-oxide catalyzed allylation and propargylation of aromatic aldehydes. Org. Lett. 2012;14:5310–5313.
[25] Sepúlveda D, Lu T, Wheeler SE. Performance of DFT methods and origin of stereoselectivity in bipyridine n,n′-dioxide catalyzed allylation and propargylation reactions. Org Biomol Chem. 2014;12:8346–8353.
[26] Rooks BJ, Wheeler SE. AARON, automated alkylation reaction optimizer for N-oxides, version 0.72. College Station, TX: Texas A&M University, 2015.
[27] Rooks BJ, Haas MR, Sepúlveda D, Lu T, Wheeler SE. Prospects for the computational design of bipyridine N,N′-dioxide catalysts for asymmetric propargylations. ACS Catal. 2015;5:272–280.
[28] Guan Y, Rooks BJ, Wheeler SE. AARON: an automated reaction optimizer for non-metal catalyzed reactions, version 0.91. College Station, TX: Texas A&M University, 2016.

[29] Yang S-C, Lankau T, Yu C-H. A theoretical study of the nornicotine-catalyzed Mannich reaction in wet solvents and water. Green Chem. 2014;16:3999–4008.

[30] Pyne SG, Au CW, Davis AS, Morgan IR, Ritthiwigrom T, Yazici A. Exploiting the borono-Mannich reaction in bioactive alkaloid synthesis. Pure Appl Chem. 2008;80(4):751–762.

[31] Dickerson TJ, Janda KM. Aqueous aldol catalysis by a nicotine metabolite. J Am Chem Soc. 2002;124:3220–3221.

[32] Zhao Y, Truhlar DG. A new local density functional for main-group thermochemistry, transition metal bonding, thermochemical kinetics, and noncovalent interactions. J Chem Phys. 2006;125:194101.

[33] Miertuš S, Scrocco E, Tomasi J. Electrostatic Interaction of a Solute with a Continuum. A Direct Utilizaion of AB Initio Molecular Potentials for the Prevision of Solvent Effects. Chem Phys. 1981;55:117–129.

[34] Lapkin AA, Peters M,, Greiner L, Chemat S, Leonhard K, Liauw MA, et al. Screening of new solvents for artemisinin extraction process using ab initio methodology. Green Chem. 2010;12:241–251.

[35] Klamt A, Eckert F, Hornig M, Beck ME, B¨Urger T. Fast solvent screening via quantum chemistry: COSMO-RS approach. J Comput Chem. 2002;23:275–281.

[36] Lei Z, Chen B, Li C. COSMO-RS modeling on the extraction of stimulant drugs from urine sample by the double actions of supercritical carbon doxide and ionic liquid. Chem Eng Sci. 2007;62:3940–3950.

[37] Lite J. Scientific American. Accessed: 15 Dec 2016. 2008. What is Artemisinin?23 DecAvailable at:http://www.scientificamerican.com/article/artemisinin-coartem-malaria-novartis/.

[38] ELSohly HN, Croom EM, El-Faraly FS, El Sherei MM. A large-scale extraction technique of artemisinin from artemisia annua. J Nat Prod. 1990;53(6):1560–1564.

[39] Lapkin A, Plucinski PK, Cutler M. Comparative assessment of technologies for extraction of artemisinin. J Nat Prod. 2006;69:1653–1664.

[40] Center for Disease Control. 2016 Dec 15 Accessed:15Dec2016. Available at: https://www.cdc.gov/niosh/pdfs/77-192c.pdf.

[41] Stevens JE, Macomber LD, Davis LW. IR spectra and vibrational modes of the hydrofluoroethers CF_3OCH_3, CF_3OCF_2H, and $CF_3OCF_2CF_2H$ and corresponding alkanes CF_3CH_3, CF_3CF_2H, and $CF_3CF_2CF_2H$. Open Phys Chem J. 2010;4:17–27.

[42] Stevens JE, Jabo Khayat RA, Radkevich O, Brown J. * CF_3CFHO vs. CH_3CH_2O: an ab initio molecular orbital study of mechanisms of decomposition and reaction with O_2. J Phys Chem. 2004;108:11354–11361.

[43] Morris W, Leung B, Hiroyasu F, Yaghi OK, Ning H, Hayashi H, et al. A combined experimental-computational investigation of carbon dioxide capture in a series of isoreticular zeolitic imidazolate frameworks. J Am Chem Soc. 2010;132:11006–11008.

[44] Qin H-L, Han J-B, Hao J-H, Kansten EA. Computational and experimental comparison of diphosphane and diene ligands in the Rh-catalysed carboxylation of organoboron compounds with CO_2. Green Chem. 2014;16:3224–3229.

[45] Darensbourg J, Mackiewicz RM, Phelps AL, Billodeaux DR. Copolymerization of CO_2 and epoxides catalyzed by metal salen complexes. Acc Chem Res. 2004;37:836.

[46] Darensbourg DJ, Yeung AD. Thermodynamics of the carbon dioxide–epoxide copolymerization and kinetics of the metal-free degradation: a computational study. Macromolecules. 2013;46:83–95.

[47] Darensbourg DJ, Bottarelli P, Andreatta JR. Inquiry into the formation of cyclic carbonates during the (Salen)CrX Catalyzed CO_2/cyclohexene oxide copolymerization process in the presence of ionic initiator. Macromolecules. 2007;40:7727.

[48] Darensbourg DJ, Yarbrough JC, Ortiz C, Fang CC. Comparative kinetic studies of the copolymerization of cyclohexene oxide and propylene oxide with carbon dioxide in the presence of chromium salen derivatives. in situ ftir measurements of copolymer vs cyclic carbonate production. J Am Chem Soc. 2003;125:7586.
[49] Darensbourg DJ, Yeung AD, Wei S-H. Base initiated depolymerization of polycarbonates to epoxide and carbon dioxide co-monomers: a computational study. Green Chem. 2013;15:1578.
[50] Darensbourg DJ, Chung W-C, Yeung AD, Luna M. Dramatic behavioral differences of the copolymerization reactions of 1,4-cyclohexadiene and 1,3-cyclohexadiene oxides with carbon dioxide. Macromolecules. 2015;48:1679–1687.

Serenity Desmond, Christian Ray and José G. Andino Martínez

5 Educational benefits of green chemistry

Abstract: In this article, we present our current state of affairs in the "greening" of general chemistry laboratories, at the University of Illinois at Urbana-Champaign. We recognize the need to quantify our environmental mark and what we plan to do to continue to strive to make our work more sustainable and educational.

Keywords: education, sustainability, laboratory

5.1 Introduction

It may seem that green chemistry's "ideal synthesis" is an unrealistic project, with high yields, no trace of pollution and cheap costs [1]. But, as difficult as it may sound, the demands to revert the direction of our current global environmental tendencies are great and it needs to start from the very core, the teaching of chemistry. Chemistry education is a sensitive task, especially when it comes to allowing students to have access to possibly dangerous chemicals. Therefore, we are confronted with providing highly specialized methods to handle dangerous substances and experiments or the possibility of offering an alternative that would involve a lower risk, but continue to enhance learning.

Changes to established chemistry laboratory curricula represent a challenge that many may not be willing to take. We consider that the operation of our General Chemistry Laboratories deserves close attention as we are convinced its magnitude has a significant environmental and social impact. The cost and level of handling the substances used in the experiments students perform during a semester is noteworthy, and we want to take this opportunity to demonstrate to the community that it is necessary and possible to use chemistry in a way that minimizes cost, is safer, and makes more sense.

We have decided to divide our analysis into three major areas:

1. Safety
2. Economic
3. Educational

As chemistry educators, it is our responsibility to constantly look for ways in which we can create a safer environment for our students and all the personnel involved in laboratory work to foster learning; however, it is important to quantify the impact of our efforts into making our laboratories "greener." We need to keep in mind that the repercussions of laboratory preparations are not only associated with the laboratory

https://doi.org/10.1515/9783110445923-005

itself but that one needs to plan carefully as starting materials may degrade overtime and safe storage of the reactive substances may not be practical. Aside from the preparation and storage, disposal of these materials or waste generated after a busy week of laboratory experiments may be difficult and expensive. With all these aspects, it is inevitable to provide alternatives that may allow us to more easily handle these tasks. A component that is normally ignored is the impact of these changes in the student's perception of the chemical process they are learning about. When the experiment instructions do not warn students about the dangers associated with the handling of a certain substance, then students should have an easier time focusing on what is important, the chemistry.

We want students to be fully aware of any measures that keep them safe and represent a low-cost alternative, which means students would learn the chemical concept the experiment is designed to teach, but also the environmental impact of conducting the experiment a particular way.

In 1998, the twelve principles of green chemistry were published [2]. These are outlined in Table 5.1.

The conceptualization of these principles laid the framework for the development of desirable objectives for chemical reactions and processes. In the last 18 years, academic laboratories began to more seriously evaluate their laboratory curriculum with respect to these objectives. The overwhelming consensus was that while "perfectly" green laboratory courses would be difficult to achieve given the fact that many of the properties that make chemicals useful are the very same properties that make them hazardous (e. g. acid/base or redox properties), measures could be taken to become more "green." Such measures could include: (a) choosing to use solvents and reagents that are less toxic and less flammable, (b) implementing "solvent-free" reactions and/or experiments, (c) considerable "scaling-down" of reaction volumes, and (d) minimizing the generation of hazardous waste.

Table 5.1: Twelve Principles of green chemistry.

1	Prevention
2	Atom economy
3	Less hazardous chemical syntheses
4	Designing safer chemicals
5	Safer solvents and auxiliaries
6	Design for energy efficiency
7	Use of renewable feedstocks
8	Reduce derivatives
9	Catalysis
10	Design for degradation
11	Real-time analysis of pollution prevention
12	Inherently safer chemistry for accident prevention

5.2 Safety

People make safety-related decisions all the time, every day. While outside the laboratory environment these decisions might seem fairly trivial such as, "Will I wear my seatbelt to drive around the block?" "Will I floss my teeth every day?" "Will I wear a helmet when riding my bicycle?", similar decisions associated with work in the laboratory can certainly have serious consequences.

Laboratory safety, while widely accepted as important, even critical to the laboratory environment, is not always in the forefront of people's minds. Often, researchers and students alike become so focused on their experiments that they consciously or subconsciously choose to make safety-adverse decisions. The Lab Safety Institute has created a Memorial Wall which lists a brief historical chronology of laboratory incidents [3].

In recent years, laboratory safety incidents have become more fatal and gained increased attention in the media. Two well-known examples in academic environments are the fatal incident at UCLA in December 2008 and the explosion at Texas Tech in January 2010. In the former incident, safety training and protocols were neglected which resulted in the air ignition of tert-butyllithium. The laboratory employee, Sheri Sangji, received second- and third-degree burns across more than 40 % of her body. She died, in the hospital, two and half weeks after the incident [4]. Following a detailed investigation, this incident resulted in the first criminal case as a result of academic laboratory safety negligence. In the latter example, at Texas Tech, the graduate students involved chose to scale-up a reaction, without obtaining the approval of their research advisor. The unsafe chemical quantity in combination with the pressure and friction involved in grinding the chemical for use resulted in an explosion which left both students seriously injured [5].

The two examples mentioned are extreme cases. Serious, or even fatal, injuries do not occur every time laboratory personnel or students choose to ignore safety; however, these examples, along with countless others, allow us to begin to understand the value of laboratory safety.

The American Chemical Society has long produced important communications such as the "Safety in Chemistry Academic Laboratories," "Guidelines for Chemical Safety in Academic Institutions," and through its Green Chemistry Institute, "Greening the Lab" and "Less is Better." These reports have progressively emphasized on the necessity to minimize safety risks. The Chemistry Department at the University of Illinois at Urbana-Champaign works closely with the Division of Research Safety (DRS) to guarantee state regulations are closely followed.

Twenty years ago, solvents such as benzene (a carcinogen), chloroform, and carbon tetrachloride were common in the organic laboratory. Today, they are largely replaced by safer chemicals. Some examples are provided in Table 5.2 [6]:

Table 5.2: Alternative Solvent Choices.

Solvent	Alternative	Explanation
Dry ice/ acetone	Dry ice/isopropanol	Isopropanol is the safer solvent to use in cooling baths as it works at about the same temperature while being less volatile, meaning a reduced risk of inhalation/exposure [7].
Hexane	Heptane, pentane	Heptane is much less toxic than hexane (which is neurotoxic) while maintaining very similar chemical properties [8].
THF	2-MeTHF	2-MeTHF is derived from renewable resources like sugarcane and corn, while THF is petroleum-based. In addition, 2-MeTHF is a cost-stable product that can also increase reaction yield while being easier to recycle [9].
	No solvent (solid state or reagents as solvents (i. e. melted reagents))	Solvent-less reactions can be used in a variety of applications. Specifically, in the organic chemistry laboratory, this is effective for aldol condensations (forming carbon–carbon bonds) [10].

The volume of chemicals used and waste generated in the academic laboratory setting is generally neglected as a significant environmental threat, and even though it tends to be smaller than that of an industrial laboratory, it still represents a major task to handle it. For our General Chemistry Labs, we generate approximately 5,000 L of waste each semester.

Ultimately, "Green Chemistry is the design and use of methods that eliminate health and environmental hazards in the manufacture and use of chemicals." [2] In considering the twelve principles of green chemistry, we now provide some of our current efforts aligning with some of these principles:

5.2.1 Prevention

As of Spring of 2016, we have removed the experiment "Wizards of the Winery II." In this experiment, students used chromic acid to promote the oxidation of ethanol found in wine samples and identified the blue-green solution containing Cr^4+ through spectrophotometry [11]. The use of chromic acid has been reduced in industry as it is a highly toxic reagent; however, the replacement of chromic acid with a more eco-friendly oxidizing agent can be challenging [12]. Paired with a revision/reduction of the organic chemistry material imparted to a second semester chemistry student, the removal of this laboratory was deemed necessary. As the first principle listed in Table 5.1 says it, "Prevention" is preferred over having to device methods to dispose of toxic waste. As the DRS states it on campus the best way to prevent an accident and reduce handling of toxic and dangerous chemicals is to eliminate the source [13].

5.2.2 Inherently safer chemistry for accident prevention

In the second semester of chemistry, we make considerable emphasis on acid–base chemistry. A common prejudice is that acids are dangerous substances and while that may be the case for strong acids, many weak acids are compounds that we find in biological scenarios or daily situations and are essential to many processes. In one of our *titration* laboratories, we use "Kool-Aid" to determine the amount of citric acid and ascorbic acid in this product. In this experiment, approximately 1 g of product is used per student. With approximately 30 students per laboratory session, 30 g of Kool-Aid are used about 15 times per week once a semester for a total of about 90 to 100 packets per semester. Needless to say, the use of this innocent product has a three-fold effect: (1) Safety is guaranteed as there is no need to prepare dilutions with concentrated acids that cost significantly more (see more below) and may degrade over time. There is no potential danger of exposing students to the substance. Even though certain rules need to be followed to be in laboratory, there are not any special precautions to handle this substance. (2) The cost is minimal. The purchase of enough material can be done at any store, stocking enough for the semester or year can be done with no risk. Handling of a solution that may contain a mixture 0.100 M citric acid and 0.100 M ascorbic acid increases the potential for accidents at different stages of the process from preparation to setting up for a laboratory session to handling by students. All that has been eliminated here and the disposal is simple. Finally, (3) students learn to recognize the presence of chemicals and chemistry in their daily lives which should make learning more effective.

5.2.3 Less hazardous chemical syntheses/use of renewable feedstocks/reduce derivatives

Our second semester of general chemistry laboratory culminates with the synthesis, purification and characterization of aspirin. While the synthesis requires acetic anhydride and sulphuric acid, the fact that the laboratory is spread in three sessions allows students to explore all aspects of a synthetic procedure with enough time to accomplish each goal successfully. After the synthesis, the product is purified and preserved for analysis. The excess of product is stored for future use, which makes this laboratory somewhat sustainable.

We are working on introducing all these measures into the teaching component for students to realize the benefit of keeping safe conditions in chemistry laboratory work. As of now, of the 5,000 L of waste produced, 10–20 % of this waste has to be disposed of by DRS, whereas 80–90 % can be either neutralized or is drain-ready upon generation.

5.3 Economic reasons

(All estimates on reagent purchasing are based on the Sigma-Aldrich catalog pricing without any discounts)

Safety improvements, especially those that require changes to infrastructure, cost money. It is also expensive to comply with all the personnel training, hazard assessments, workplace surveillance, medical evaluations, record keeping, etc. Certainly, large projects are costly, although in many cases they are necessary. In our chemistry department, we recently completed the renovation of one of our main buildings, the "Chemistry Annex." The Chemistry Annex was originally built in 1930 [14] and reached a point when its operation was not sustainable, running it represented a significant waste of money and energy. With a budget of $24.9 million [15], the renovation completely remodelled the building, removed aged laboratory benches and equipment as well as improved the energy efficiency of the building. The laboratories are state-of-the-art facilities with modern laboratory tables. As an example, ventilated oval air stations which occupy a smaller space than a hood were designed for use in the introductory chemistry laboratories for more efficient trapping of fumes. The "Chemistry Annex" is on its way to receive the certification of "green building" by the Leadership in Energy and Environmental Design organization [16], and it is projected to represent significant savings for years to come.

Cost of operation and materials is probably the strongest motivator to implement changes to chemical processes and experiments even over safety and toxicity. From an industrial standpoint, The Toxic Substances Control Act (1976) regulates the 80,000 chemicals in industrial use, but does not require that manufacturers test the toxicity of chemicals before they are used in commerce [17]. This, therefore, results in much of the design, selection and use of chemicals in the United States to be driven by history, function, price, rather than safety and toxicity. The purchasing and/or preparation of complex and reactive chemicals for laboratory experiments are an important and constant responsibility in a university chemistry department.

In the next paragraphs, we want to elaborate on the financial aspect of some of the changes mentioned in the previous section. We consider this to be a great beginning to shaping our introductory chemistry laboratories to have a more significant "green" component than they have had up to this point.

5.3.1 What are the costs that go into our laboratories?

From our position as facilitators: chemicals, waste disposal, materials (disposable – paper towels, paper, markers, etc. and not – glassware, which is supplied by a student laboratory fee), staff (both preparation assistants and teaching assistants (TAs)), and overhead for equipment/systems, time. It is always a matter of priority, the question of where does it make sense to try to save money? We are clear that we do not want to compromise our academic goals of providing students with a strong foundation of chemical principles, but we are finding that changing some of our current experiments and practices may even enhance our programme.

Even though equipment such as analytical instruments represents a significant expense, they are hassle free for several years due to warranty and the fact that they should last long enough before we may find the need to replace a substantial number of units. We will focus on other expenses and how they are associated with turning into "green" practices.

The easiest expense to cut down is the purchase of chemicals. If we decide to remove an experiment that uses an expensive chemical and replace it with a cheaper alternative, there are savings associated with the lower cost of the purchase and also the need or not of disposal of the post-experiment excess and by-products.

I. Aside from the safety benefit of not using the "chromic acid reagent" for the alcohol oxidation experiment, the elimination of it represents savings of between $10,000 to $15,000 per semester. This is only associated with reagent purchasing, and an additional amount would be incurred on the special disposal of excess reagent and by-product waste. It is clear that this change is convenient due to changes in our general chemistry curriculum, but it is an important financial gain nonetheless.

II. As we had mentioned above, the purchase of Kool-Aid over chemical reagents to teach acid–base titrations has signified a substantial cut in costs. The initial cost of the reagent approximates $100 per semester, which can be considered negligible. The cost of the production of deionized water need not to be considered as we would have to provide it under any circumstances. As far as the disposal is concerned, according to the DRS any solution within a pH of 6–10 can be disposed of down the drain with enough running water; therefore, the disposal of the excess and by-products from this practice may or not even require a simple neutralization prior to disposal. There is no requirement for storage or picking up of sensitive waste.

Our goal is to be able to provide similar alternatives to at least 50 % of our general chemistry experiments within the next 5 years.

5.3.2 What are the costs that students incur from our laboratories?

We do not want to be oblivious to the expense students acquire when participating in our chemistry laboratories. Up until the fall of this year (2016), students have needed to purchase a chemistry laboratory notebook and the chemistry laboratory manual and are charged a usage fee, totaling approximately $50 per semester. Notwithstanding, for over 5 years, we have created and provided a series of online experiment companion assignments through the Learning Online Network with Computer-Assisted Personalized Approach platform or LON-CAPA [18]. Access to LON-CAPA is provided to the students free of charge. An important component of our laboratory work evaluation is the completion of online prelab, laboratory and postlab assignments through LON-CAPA. Students are able to prepare for laboratories, enter data during laboratories and complete a report-like assignment in which they address questions that allow students to think about the chemistry they learned while conducting the experiment.

Keep in mind that all of this is done without the use of a single piece of paper, and it is graded automatically without the need to depend on TAs returning graded reports. Students are still required to provide copies of the pages that contain their work in the laboratory notebook to the TA, but these do not count for any grade and are only used if we need to confirm the presence of a student on a particular date. This made us wonder about the use of the laboratory notebook. A problem that we noticed was that some students did not use the notebook as expected, meaning that they prepared very poor prelab procedures, with very limited and unclear information or none at all. While this constitutes a deficiency in the learning of some of the principles we expect students to take with them from the laboraotry, it also represents a significant waste of money and paper if the notebooks will not be used as expected during and after the semester has ended.

Beginning in 2017, the purchase of a chemistry laboratory notebook will be optional for the second semester of general chemistry students. Instead, they will be provided with a couple of pages at the end of the experiment description and procedure in the laboratory manual to write a procedure, equation and tables. The new manual will cost around $35. This represents around $10,000 in savings for the students and around 550 kg of paper that will not be wasted.

Finally, our multi-week laboratory dealing with the synthesis and subsequent characterization of aspirin uses products students have created and saves reagent costs/preparation for the following week.

We consider this an important step into trying to reduce chemical waste as well as materials, but also resources that can be redistributed to cover other areas and or create new more sustainable experiments.

5.4 Educational advantages

"DON'T CHANGE THE EDUCATIONAL GOALS. THESE NEED TO BE PRIORITY #1!"

Our general chemistry laboratories are intended to provide students from different backgrounds with the fundamentals of chemistry in a curiosity-stimulating environment. The academic laboratory is the perfect platform within which to begin to incorporate green techniques and principles. It is here that the student can learn these methods, incorporate them into their laboratory practices and then be better prepared when they enter the chemical workforce [19].

It seems to be the case for many students that, when they come to a chemistry class at the undergraduate level, they are intimidated by a variety of reasons: a poor high school chemistry experience, the lack of understanding of chemical formulas, the misconception that chemistry is dangerous or even the fear to comply with the demands of the course itself [20]. This state of uncertainty is the worst scenario to promote learning. Let's face it, students tend to be almost predisposed to hate chemistry, but anecdotally they appear to be more attracted to learn to conduct chemistry experiments. This makes the chemistry laboratory a bridge that connects us to the students, even when they may have a hard time seeing the link with the theory.

As we have seen above, safety and cost are a major push to discontinue certain experiments and adopt new ones that use less toxic and/or expensive materials. As we serve a diverse pool of students, many of whom are not going to pursue a chemistry degree, the use of complex chemicals is probably a not very effective strategy to engage students and it is bound to leave many students even more confused. We consider that the use of common substances allows students to focus on the science and removes the stress. By eliminating toxic and expensive substances and changing to more benign alternatives, we enhance the education of our students, not only with a strong chemistry foundation but also on the impact of chemistry in everyday society.

In 2012, a C&EN News article [21] noted that the implementation of the "Green Chemistry Commitment" gave "Green Chemistry" a prescriptive facet that it was not intended to have. Although we now know the "Green Chemistry Commitment" exists to provide resources and a community that works together to push for faster progress in the field, we must not get lost in semantics and get to work one experiment at a time. However, it is no surprise that the problem with sustainability and deterioration of our environment is not only a matter of discussion for chemists and physical scientists. The situation is so serious that in recent years the term "Green Chemistry" has transcended the chemistry realm and has become a topic in educational and social panels worldwide [22].

We want to take a very pragmatic position about what the changes to our curricula and laboratory experiments involve and how they can make a significant difference for our students' learning process. For example, in teaching acid–base chemistry, the notion that acids are dangerous may distract students from focusing on the important aspects of acid–base chemistry. We believe that by using Kool-Aid we allow students to think more easily about the chemistry. Using Kool-Aid instead of other acid solutions provides multiple advantages. Probably, all students have consumed Kool-Aid. They are not afraid of it, they know it won't hurt them, but curiously there are many aspects of it that they have no idea about. In fact, the discovery of the "secrets" of Kool-Aid sparks an amazing desire to discover more about the substance and maybe other household substances.

These changes put the students in control to start making important decisions to create judgements about chemistry, which is what we ultimately would want them to do. This may even inspire students to pursue a degree in chemistry. Once they are confident about the fundamentals, they will be able to tackle more complex systems.

As mentioned in the previous section, we use LON-CAPA to manage our laboratory courses. We would like to take this approach to the next level and rely more heavily on assessment through this online tool. How do we assess the laboraotry?

Depending on the educational goal of the laboratory, we need to evaluate our assessment. For some laboratories, learning how to write a laboratory report or learning how to write journal-style papers is important. However, grading these assessments is very time consuming and the rigor in grading depends on the TA. The chemistry major's series of introductory laboratories are the only ones in which students need to write one part of a report for each laboratory; they learn how to make a full report at the end, but save time/writing along the way.

Students who are not in a chemistry track may or may not benefit from a rigorous report-writing curriculum. If we are to only use LON-CAPA to provide assignments and expect students to obtain the foundation that will tell us they are prepared and our academic goals have been met, then the material that they receive must be of the highest quality, which becomes challenging at times if we have to circumvent a technological problem.

A valid question centers around the validity of the online assignment method using LON-CAPA and how bulletproof is it with respect to academic dishonesty. To reduce or even eliminate the potential for academic dishonesty with the online data submission during the course of a laboraotry, we have been able to create a dedicated online network that prevents students from signing into the laboratory assignment while the laboratory experiment is taking place from areas physically remote from the laboratory room. This virtually eliminates the need for any paper evidence of the student's work. We realize that, however, this is still contingent upon the particular instructor's preferences, although we plan to evaluate the possibility of making it a general policy as it aligns us with a less wasteful laboraotry culture.

Last, we have, over the last few years, identified some weak points in the conduction of general chemistry laboratory courses. Some of the most critical points are:

a. Students not complying with safety rules
b. Not completing data submission during laboraotry resulting in inability to do the postlab
c. Reliability on TAs to help to apply safety rules and be knowledgeable about the upcoming experiment.

During 2016, a series of measures implemented by the Laboratory Coordination team with the use of our online tools has seen a great improvement in all respects. Regular communication via LON-CAPA, providing audio-visuals that describe clearly what the students need to do to comply with the regulations has made a big difference and has substantially reduced the number of students that do not want to adhere to the protocol. This makes it much easier to identify the students that may need to be asked to leave the laboraotry to correct their fault. Also, a rigorous TA training programme, similar to others found in the literature [23], responds to the necessity to have reliable TAs. As of 2016, we have had TA training sessions and TA meetings for each experiment which helps tremendously with the problem of students getting the wrong advice. When TAs know the details of the experiment, they are motivated and their interaction with the students is much more fruitful. Additionally, TAs encourage and evaluate if students have left their stations clean, to prevent any safety problem with the next group and in return students could collect a minimum amount of points which they receive gladly. This system has also been extremely successful and has made a great difference in the condition in which the laboratory is found at all times. The problem of not completing data submission has been basically eliminated as our students are now required to stay the complete laboratory period regardless of them finishing data submission. They are then required to continue on with the postlab assignment, which is due a week later. This way students do not lose points unnecessarily rushing out of the laboratory. All these changes have been facilitated by the use of LON-CAPA, and we plan to create more tools that allow us to maintain a consistent pace to offer students a rich variety of chemistry experiments with an emphasis on safety and sustainability.

We are sure that as these changes are yielding excellent results internally for our purpose of conducting the laboratory, they will also be shaping the way students carry themselves in their future challenges.

5.5 Conclusions and future work

In conclusion, we present here a group of basic but steady changes that we deem necessary to be able to teach students the chemistry fundamentals that will prepare them for their chosen disciplines. Also we are beginning to make a clear note of the environmental benefits of conducting various laboratories with environmentally benign alternatives and saving materials that will also teach students the importance of sustainability. We plan to offer more experiments with Green Chemistry alternatives by scaling down where appropriate and trying to implement solvent-less experiments, e. g. caffeine extraction with supercritical CO_2. We are currently creating assessment tools that will allows us to compile the data necessary to measure the impact of all these implementations and create a more informed and scientific description of our success.

References

[1] Clark JH. Green chemistry: challenges and opportunities. Green Chem. 1999;1(1):1–8.
[2] Anastas PT, Warner JC. Green chemistry: theory and practice. New York: Oxford University Press, 1998:30.
[3] XXXX labsafetyinstitute.org/memorialwall.html.
[4] Kemsley J. Learning from UCLA. Chem Eng News. 2009;87(31):29–31 3334.
[5] Kemsley J. Texas tech lessons. Chem Eng News. 2010;88(34):34–37.
[6] ACS. Greening the lab and beyond!, 2014.
[7] XXXX http://www.chem.utoronto.ca/green/_shared/pdfs/Simple%20Techniques%20to%20Make%20Everyd ay%20Lab%20Work%20Greener.pdf.

[8] Takeuchi Y, Ono Y, Hisanaga N, Kitoh J, Suguira Y. A comparative study on the neurotoxicity of n-pentane, n-hexane, and n-heptane in the rat. Brit J Ind Med. 1980;37:241–247.

[9] Aycock DF. Solvent Applications of 2-Methyltetrahydrofuran in Organometallic and Biphasic Reactions. Org Process Res Dev. 2007;11:156–159.

[10] Doxsee KM, Hutchinson JE. Green organic chemistry: strategies, tools, and laboratory experiments. Belmont, California: Brooks/Cole Cengage Learning: University of Oregon, 2004.

[11] Department of Chemistry, University of Illinois at Urbana-Champaign. Chemistry 105 - General Chemistry Experiments. Fall. 2015;96.

[12] Karimipour GR, Montazerozohori M, Karami B. Tungstate sulphuric acid/CrO3 as a novel heterogeneous system for oxidation of alcohols to carbonyl compounds. Oxid Commun. 2011;34(3):622–626.

[13] XXXX https://www.drs.illinois.edu/Waste/ChemicalWasteCollectionAndStorage.

[14] XXXX http://www.trustees.uillinois.edu/trustees/agenda/March-6-2014/017-mar-Chem-Annex-Renovations.pdf.

[15] XXXX http://www.usgbc.org/projects/uiuc-chem-annex-renovation-and-addition.

[16] XXXX https://www.epa.gov/laws-regulations/summary-toxic-substances-control-act.

[17] XXXX http://www.lon-capa.org/.

[18] Hjeresen DL, Boese JM, Schutt DL. Green chemistry and education. J Chem Educ. 2000; 77(12):1543.

[19] Silberman RG. Problems with chemistry problems: student perception and suggestions. J Chem Educ. 1981;58(12):1036.

[20] Ritter S. Teaching green. Chem Eng News. 2012;90(40):64–65.

[21] Bodlalo LH, Sabbaghan M, Jome SMRE. A comparative study in green chemistry education curriculum in America and China. Procedia Soc Behav Sci. 2013;90:288–292.

[22] Dragisich V, Keller V, Zhao M. An intensive training program for effective teaching assistants in chemistry. J Chem Educ. 2016;93(7):1204–1210.

[23] XXXX http://uihistories.library.illinois.edu/virtualtour/maincampus/chemannex/.

Daniel Y. Pharr

6 Green analytical chemistry – the use of surfactants as a replacement of organic solvents in spectroscopy

Abstract: This chapter gives an introduction to the many practical uses of surfactants in analytical chemistry in replacing organic solvents to achieve greener chemistry. Taking a holistic approach, it covers some background of surfactants as chemical solvents, their properties and as green chemicals, including their environmental effects. The achievements of green analytical chemistry with micellar systems are reviewed in all the major areas of analytical chemistry where these reagents have been found to be useful.

Keywords: surfactants, green chemistry, micellar solutions

6.1 Introduction

A review of micelles (i. e. surfactants, detergents, and soaps) as green chemical reagents seems to be an almost endless task as new results are constantly found in the literature. The effectiveness and properties of micelles will be illustrated through examples with numerous references to other articles and reviews that readers can explore on their own. Rather than give a complete review, which might have to be encyclopedic in nature, one hopes to give enough background that readers can investigate further and that they can understand the utility of surfactant systems as green chemical reagents. Because surfactants have been studied for an extended period, these mature chemical's properties are well known, and with their widespread use in industry their safety and cost can be optimized.

6.2 General aspects of surfactants

In the discussion of surfactants, chemists often focus exclusively on the surfactant molecules and/or their relationship with the chemical(s) with which they react. Before we start that dialogue, briefly consider the surfactants' solvent: water. If water were not so abundant, we would marvel at its unique properties. Although only 18 g/mole, the temperature range at which water is in the liquid state is much broader and much higher than would normally be expected. This phenomenon is of course due to hydrogen bonding illustrated in Figure 6.1. If one thinks of the crystal structure of a diamond with each tetrahedral carbon bonded to another, one can

https://doi.org/10.1515/9783110445923-006

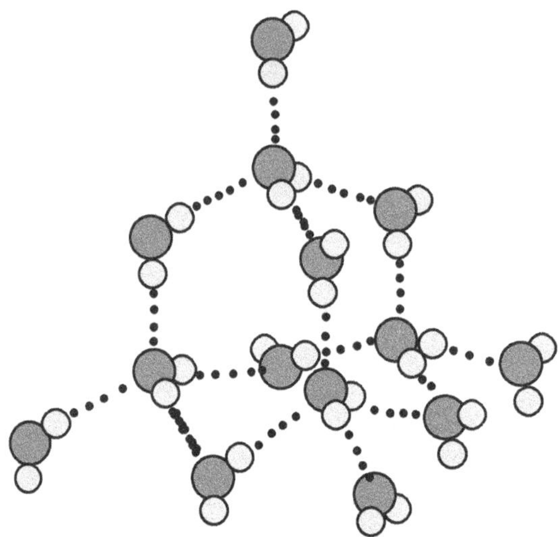

Figure 6.1: The diamond shape of pure water's hydrogen bonding.

imagine why this macro-molecule is so hard; however, extremely pure water has a similar structure to a diamond since each molecule is connected to another through a similar tetrahedral matrix, this time held together by hydrogen bonding. Although not as strong as a covalent bond, this hydrogen bonding gives water its large liquid temperature range and cause it to be, in its pure form, a non-conducting liquid.

Adding other chemicals to water breaks up this diamond-like matrix and increases the entropy of the liquid, breaking up an ordered state into a more disordered state. The composition of these additional chemicals has a direct effect on the physical properties of the water–chemical mixture. For example, a soap surfactant is like two chemicals in one: a salt and an alkane. If the alkane length is short, then the surfactant is more soluble in water like ethanol compared to octanol. The salt functional group with an anionic charge can in the simplest terms be seen as a carbonate group. So, when one increases the alkaline earth concentration, one obtains a precipitate similar to if $MgCO_3$, $CaCO_3$ $SrCO_3$, or $BaCO_3$ were added to water. To think of it another way, in a separatory funnel with an organic layer and a cloudy aqueous layer, adding a soluble salt will clear the aqueous phase because the dielectric nature of water will increase with the salt and push the organic alkane out of the aqueous phase. This same reaction occurs when a micelle forms in water, lowering of the molarity where micelles form if a salt is added to an aqueous surfactant system [1].

The earliest version of soap may date back to Babylonian times, with soap still today the most used surfactant world-wide but has been replaced in areas with hard water and in developed countries by synthetic anionic surfactants. Several financial reports estimate that surfactants and detergents are an over 30-billion-dollar business. Their uses have been more varied than one might originally expect: during the

Great War, Germany started using detergents due to a shortage of fats needed for the war effort because of the Allied blockade. More recently, chemists first considered surfactants as chemical reagents due to their useful major chemical properties. Surfactant chemistry has been so widespread due to their ability to separate nonpolar substances in water, but surfactants may also act as wetting agents (they lower the surface tension of water), emulsifiers, foaming agents, and dispersants, and have been used in aqueous and non-aqueous systems [2–4]. With the advent of Green Chemistry, their environmental friendly aspect has added a new reason to consider replacing organic solvents with surfactants. Much work has now been done in this field. For example, surfactants' potential contribution to Green Chemistry is discussed in a 2001 review by Urata [5]. The type and amount of organic waste is tremendously reduced using aqueous surfactants systems that only contain 10^{-2} to 10^{-4} M surfactant in water compared to almost 100 % organic solvent.

Water, the most popular of solvents, will not dissolve molecules and complexes that are non-polar in nature, however, when mixed with surface active agents, (surfactants), oils and grease can be washed away. The solubilization occurs because the surfactants can form micelles, which are created at what is termed the critical micelle concentration (CMC). Below this value ionic surfactants act like electrolytes. The number of molecules needed to form an individual micelle is called the aggregation number. In water a dynamic three-dimensional ellipsoid or sphere occurs at the CMC as the hydrophilic head groups positioned outward in the aqueous solution and the hydrophobic long chain hydrocarbon are facing toward a center of the micelle, whose general structure is shown in Figure 6.2 [6]. The Stern layer consists of the

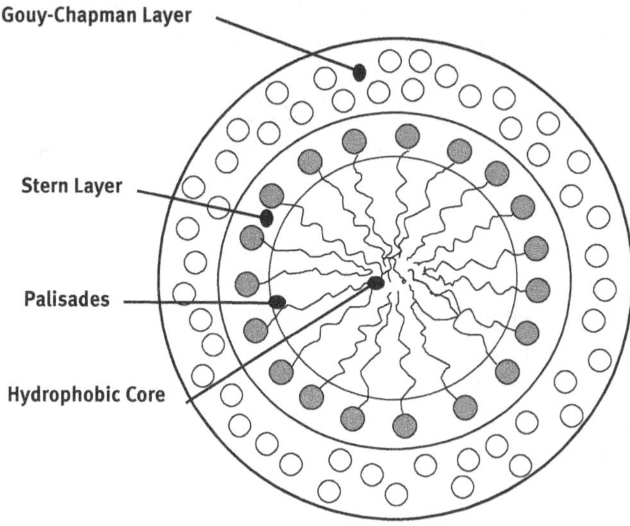

Figure 6.2: Micelle.

hydrophilic groups that face outward toward the bulk water. The region adjacent to the Stern layer that contains a high density of counter-ions of the polar heads (Gouy–Chapman double layer) separates the hydrophobic interior of the micelle from the bulk aqueous phase outside. The center space is where the non-polar organic compounds can be solubilized. Fluorescence polarization experiments determined that the core of the micelle is hydrocarbon-like and fluid, microviscosities of 10 to 30 cP having been observed. The nature of the core has been extensively studied using Raman spectroscopy, fluorescence probes, positron annihilation and sound velocity measurements and excimer formation [7] and temperature dependence [8]. Micelles are distorted spheres, ellipsoidal in shape of small dimensions, with 1.5–3.0 nm radius. At higher surfactant concentrations much larger micelles may be formed which are rod like in character. Micelles are dynamic in nature with movement of the counter-ions and water on its surface and with the exchange of surfactant molecules from the micelle to the bulk solution. Generally, the type of polar head group indicates one of five surfactant types: nonionic, cationic, anionic, zwitterionic and gemini. Each of these groups has many examples and classes of surfactants. Examples of some common surfactants and their characteristics are given in Table 6.2.

As an anionic surfactant, soap normally has a carboxylate group from the saponification of a fat by sodium hydroxide, but other anionic groups are possible: sulfate, sulfonate, dioctyl sulfosuccinate and phosphate. For laundry detergents the classes are alkylbenzene sulfonates, alkyl sulfates and alkyl ether sulfates, which are less susceptible to hard water but foam more than other detergents, secondary alkane sulfonates and soap.

For cationic surfactants, there are octenidine dihydrochloride, a variety of quaternary ammonium salts, esterified mono- or di-alkyl quaternary compounds, esterquats, and imidazoline derivatives. One of the first to achieve widespread use as a fabric softener was *N,N*-Dimethyl-*N*-octadecyloctadecan-1-aminium chloride, which has been phased out because of its low biodegradability [9]. This is an example of how the use of a mature industrial chemical can be chosen with a clear knowledge of its environmental properties compared to chemicals only used in academic research and they have not been fully studied.

The quaternary ammonium salt can also be the positive part of the zwitterionic (amphoteric) surfactant with the negative part containing a variety of functional groups such as sulfonates, phosphates (often found in nature), carboxylates and others. These surfactants are not used as widely because of their higher costs compared to other products. Zwitterionic surfactants include hydroxysultaines, alkyl betaine, and alkylamidopropyl betaine. Betaines are derived from imidazolines and alkylamphoacetates. These surfactants are constructed on salts of quaternary ammonium cations where the fatty acid is linked to the center of the molecule through ester linkages. This type of zwitterionic surfactant is commonly referred to as betaine-esters or esterquats and is more biodegradable than the early cationic surfactants that were also used as fabric softeners [9].

Gemini surfactants are dimeric surfactants with a spacer in between. These molecules have two hydrophilic groups and two tails per surfactant molecule. Gemini surfactants enjoy a number of superior properties when contrasted to the other single-headed, single-tailed surfactants. They exhibit smaller CMC values, have increased surface activity and less surface tension at the CMC, are more hard-water tolerant, and have superior wetting times as well as lower Krafft points. There have been several reviews of both cationic and anionic Gemini surfactants [10–13].

There are several types of nonionic surfactants: polyethylene glycol alkyl ethers or alcohol ethoxylates, polyethylene glycol alkylphenyl ethers, fatty acid alkanolamides, alkylamine oxides, *N*-methylglucamides, polyoxyethylene glycol sorbitan alkyl esters and alkylpolyglycosides. The popular type that has polyoxyethylenated head group forms hydrated coils in the outer region of the micelle that act as the hydrophilic part of the micelle. A study of nonionic surfactants using Raman spectroscopy reported additional Raman lines were seen. These Raman peaks indicate the liquid-like nature of the micellar core. The Raman spectroscopy also denotes that the long ethylene oxide chains in some nonionic surfactants assume dihedral helical structures, typical in lengthy molecules. However, the nonionic surfactants with shorter ethylene oxide chains (Triton X-100) have a major portion of the ethylene oxide chain in an open coil structure (not a dihedral helical structure) and are hydrated instead of coiling on themselves [14].

6.2.1 Determination of the CMC

Numerous methods have been used to determine a surfactant's CMC including spectroscopy, UV/Visible, IR, fluorescence, nuclear magnetic resonance, electrode potential/conductivity, voltammetry, also scattering techniques, calorimetry, surface tension, and foaming. This analysis is generally accomplished by plotting the surfactant's concentration versus the physical property under investigation. Higher temperatures and pressure can make the CMC difficult to determine because the CMC may appear to occur at a wider range than would be the case in ambient conditions. Goodling et al. have characterized the CMC and aggregation number for sodium dodecylsulfate, SDS, by luminescence in an undergraduate experiment [15]. In a more recent study by Nakahara et al., the CMC values for both nonionic and anionic surfactants were assessed employing a photosensitive monoazacryptand–barium complex. Its fluorescence intensity is perceptively altered by environmental conditions established on the photo-induced electron transfer mechanism as a fluorescent probe, and this result was compared to the CMC's previously reported in the literature that used 1,6-diphenyl-1,3,5-hexatriene, or 8-anilinonaphthalenesulfonic acid magnesium salt, or pyrene, as probes [16]. Besides being used to determine the CMC, luminescence has also been used to determine other surfactant characteristics:

shape, size, aggregation number and microviscosity using not only fluorescence intensity but also fluorescence lifetime, anisotropy and quenching measurements [17]. The change in the added fluorescent dye's color could be solely due to the change in the microenvironment from the formation of micelles at the CMC but other factors may also play a role. The dye may interact with the surfactant in the formation of the micelle, thereby altering the aggregation of the micelle. The dye may form an ion-pair or close ion-pair with the surfactant or a dimer with itself or form a special kind of micelle (mixed micelles) at concentrations far below the normal CMC characteristic of the surfactant as reported by Garcia et al. [18] These researchers further propose that the interaction of the surfactant with a $-SO_3^-$ promotes electron withdrawing on the aromatic dye. Thus, with a lowering of the pKa of any –OH group, there is an ionization of easily dissociated groups and a change in the chromophore's structure that is seen spectrophotometrically [18].

Electrolytes added to ionic surfactant solutions exhibit a lowering of the CMC, which has a linear dependence of log (CMC) on the concentration of added salt [19]. This phenomenon is not true for nonionic surfactants and their CMC values. When non-electrolytes are mixed with the surfactant media, the effects are reliant on the nature of the species added [18]. For polar non-electrolytes (e. g., *n*-alcohols), the CMC diminishes with increasing concentration of alcohol, whereas when urea is added to micellar solutions, this addition tends to increase the CMC and may even prevent micelle formation. Nonpolar additives are apt to have slight effects on the CMC. Usually, the occurrence of 20–40 % (v/v) of organic co-solvents (ethanol and acetonitrile) in water prevents micelle formation resulting in a reduction of fluorescence intensity [20].

6.2.2 The Krafft and cloud points

These two unique surfactant characteristics can be used for the effective separation of an analyte, which will be discussed later. The Krafft point is the minimum temperature at which surfactants form micelles. Below this value the surfactant remains in crystalline form, even in an aqueous solution. Above the Krafft point a fairly large amount of the surfactant can be dispersed in micelles, and solubility intensifies. Gu and Sjöblom have reviewed a series of surfactants and found that there is a linear relationship between the Krafft point and the logarithm of CMC for ionic surfactants [21]. These relationships have a continuous negative slope for different kinds of homologues of surfactants with the same kind of counter ion. For homologues of nonionic surfactants, a comparable linearity is seen; however, these surfactants exhibit a positive slope. Apparently this phenomenon reveals the subtle balance between the attraction and repulsion forces of surfactants, accountable both for phase separation and micelle formation [21].

The cloud point, the temperature where the mixture becomes cloudy as there is a phase separation and two phases appear, can occur with nonionic surfactants who do not exhibit Krafft points. Hinze published two excellent reviews on cloud point extractions (CPEs) in 1993 with Pramauro [22] and again in 1999 with Quina [23].

6.2.3 Probing the micelle environment

Several probe molecules that show altered spectroscopic behavior based on their environment have been used to study micelles. These investigations provide worthwhile information on the nature of diverse regions of micelles, that is, their degree of rigidity and polarity in the core and on the surface regions. The fluorescence emission spectra fine structure of pyrene changes with its solvent environment [24]. Specifically, the ratio of the third fluorescence peak to the first peak has been used in probe studies of micellar systems to compare them with both water and organic solvents like hexane and methanol [25]. This III/I ratio increases distinctly on altering the solvent from water to sodium dodecylsulfate, SDS, although the ratio was smaller than that detected in dodecane. This ratio change shows that the environment experienced by pyrene is somewhere between that of an alkane and water. The addition of pentanol which will go into the head group region of the micelle and thereby push the pyrene into the micelle core, this effect is seen by III/I ratio increase indicating it is in an organic environment [7]. Pyrene was used as a probe to determine the hydrophobicity of microenvironments and the partition coefficient in *n*-β-octylglucoside micelles [26].

Pyrene has also been used in the fluorescence quenching studies of SDS and CTAB micellar systems by changing the type and concentration of the counter ions and upon the addition of neutral molecules like benzyl alcohol [27, 28]. From the spectroscopic data of the polar molecule benzophenone (also acetophone and pyridinium ions) whose absorption spectrum is solvent dependent, one can conclude that this molecule is located on the micellar surface rather than the core. These molecules can also be employed to provide a measure of degree of local polarity for the surface of the micelle. Benzene and naphthalene were reported to have a hydrocarbon-like environment in micelles based on NMR studies [25]. A plot of the solvent's dielectric constant versus wavelength maximum exhibited a line that showed a red shift with increased ε, molar absorptivity [29]. Drummond et al. used 2,6-diphenyl-4-(2,4,6-triphenyl-1-pyridinio)phenoxide {Reichardt's dye, CAS No. 10081–39–7} to compile a table of effective interfacial dielectric constants in organized media and to measure the electrostatic surface potential for a series of cationic surfactants [30]. The explanation of this trend comes from the two differences in the same molecule: the aromaticity and the aldehyde group. In a nonpolar environment the fluorescence maximum is 400 nm (this is due to an n–π^* transition). When the polarity of the micro-environment is increased so is the wavelength maximum, whereby the π–π^*

level (which lies close to the n–π * level), is brought below that of the n–π * by solvent interaction with the excited state [25]. The effect of quenching on the probe molecule was also investigated to further elucidate the nature of the micelle [25]. Prodan, 6-propionyl-2-(dimethylamino)naphthalene {CAS No. 70504-01-7} was used to determine the aggregation number of SDS micelles and showed that the aggregation number changed in the presence of sodium chloride: 91 (0.1 M NaCl), 105 (0.2 M NaCl and 129 (0.4 M NaCl), and several other interesting characteristics of micellar behavior as its excitation and emission wavelengths vary depending on its microenvironment [31]. 1-Naphthol {CAS No. 90-15-3} was used as a probe to study the difference between three different micellar systems: anionic–SDS, cationic–CTAB and nonionic–Triton X-100R [32]. In a study by Almgren et al. the kinetic equilibria between small neutral arenes and various ionic surfactants as the exit and reentry of the molecules in and out of the micelle were measured quantitatively using phosphorescence as a monitor for the processes in different deoxygenated aqueous micellar solutions and residence times of 1–100 microseconds were reported [33]. Conversely, a micelle will, typically, lose a monomer in 0.01–10 microseconds, so some monomers will be exchanged, during the time that the phosphorescence probe resides in the micelle. They also studied the solubility trends of some eleven aromatic compounds and found that in general that the solubility increases with increasing chain length of the hydrophobic portion of the surfactant with a noticeable difference between the cationic surfactants and the anionic surfactants studied. Their conclusion was that pyrene is preferentially solubilized at the surface of alkylammonium–surfactant micelles. Three different indoles (indole, 1-methylindole and 3-methylindole) were used to study the effects that Brij-35 had on their fluorescence since their fluorescence attributes in a different environment permits their association with the micelle to be quantified. In particular control of the odor caused by animal waste from 3-methylindole was address as being controlled and measured in surfactant systems [34]. This is seen as a possible solution to waste management.

In a study by Martens and Verhoeven, a SDS micellar solution exhibited a dramatic enhancement in the amount of ground-state electron donor–acceptor complexation between pyrene (donor) and *N,N′*-dimethyl-4,4′-bipyridinium dichloride, pq^{2+}, (acceptor) when compared to homogeneous solutions or CTAB. The charge transfer absorptions observed evidence of a direct contact of the two species. Previous work hypothesized only a dynamic mechanism, but this work shows that a static mechanism also contributes to the overall mechanism in SDS micellar media because the pq^{2+} is held on the anionic surface of the micelle while the pyrene is in the interior facilitating the reaction by proximity [35]. A study of the fluorescence quenching effects of methylene iodide or nitromethane on aromatic compounds in the presence of both SDS and copper (II) ions with anthracene in CTAB were in agreement with the estimates of the kinetic equations and yield values for exit and entry rates of guest molecules in the micelle; the study demonstrated that the quencher is not disturbed by the excitation of the probe [36]. In a *Journal of*

Chemical Education article, a kinetics demonstration shows the effect of micellar stabilization of excited state deprotonation of the 8-hydroxypyrene-1,3,6-trisulfonate ion {CAS No. 6358–69–6} being stabilized by the cationic surfactant CTAB [37].

These probe studies lead plainly to the direct conclusion that the movement of molecules in the micellar interior is affected by the micellar size, charge, counter ions, shape and viscosity regardless of whether the micelle is a spherical or rod-like shape. A review by Grieser et al. examines the many different spectral probe techniques that have been used to study micellar and vesicular systems [38].

6.2.4 Catalysis

Besides solubilization, surfactants have been used in catalysis, photolysis and extractions. The hydrophobic tail is most often a long chained hydrocarbon (linear, branched or aromatic) but it may consist of a fluorocarbon, a siloxane or even a double tail. The ability of organic compounds to be dissolved in water with the use of micelles can lead to the catalysis of the reaction between the reagents and the analyte. The pseudophase ion-exchange model proposes that reactions are primarily affected due to the proximity of reactants in a small micellar volume [39]. For example, when the cationic micelle attracts the anionic analyte, it facilitates a faster kinetic reaction by its proximity to the organic reagent inside the micelle. The reaction is no longer based on random collisions in a solution, as the reagents are closer together and in the proper orientation for the reaction to occur. For some organic reagents there are a combination of several factors that affect its solubility and reactivity in a micellar solution. The nonpolar solute may exhibit a deep penetration or a short penetration of the core at a shorter distance from the Stern layer. It is also possible that the ionic solutes or ionic portions of an organic solute may be absorbed or repelled by the polar micellar surface. This charge may also affect the kinetics of the desired chemical reaction. These effects entail the dynamics of the hydrophobic and the electrostatic interactions occurring in the micellar system with the reactants. Because of the different properties of each surfactant's polar end, different surfactants are expected to behave differently as has been reviewed [7].

A kinetic study of the rate of fading of triphenylmethane dyes and of sulfonphthalein indicators were determined in alkaline solution in the presence of micelle-forming surfactants. The kinetics of a cationic dye like crystal violet was significantly accelerated by the addition of CTAB, and retarded by SDS but not as great effects were seen with the sulfonphthalein dyes [40]. Reeves studied the kinetics and nature of the aggregates (dyes complexing with or dimerizing with the surfactant) of the base-catalyzed hydrolysis of acetate and hexanoate esters of an azonaphthol sulfonate dye in the presence of two different cationic surfactants at various concentrations [41]. This study was instigated to help explore the different aggregation numbers determined when different dyes were used. In a companion study Reeves investigated the effects of diverse counter-ions and various concentrations on the base-catalyzed hydrolysis reaction of hexanoate esters of an azonaphthol sulfonate dye exhibited not only absorption wavelength shifts but also changes depending upon the CTAB concentration and the method of making the original species mixtures [42].

For example, for the hydrolysis of cytotoxic pyronins a threefold higher rate constant in an aqueous solution of cationic CTAB while the hydrolysis reaction was completely inhibited by anionic SDS [20]. Another example is the fluorescence analysis of cyanide ion using its reaction with 1,4-naphthaquinone-2-sulfonic acid; in the presence of CTAB micelles, the reaction takes 5 min as opposed to a reaction time of about 90 min without CTAB [20]. A most dramatic increase of 10,000 times is seen in the fluorescence analysis of thiols with 4-nitro-*N*-*n*-butyl-1,8-naphthalimide using a CTAC surfactant system [43].

A series of studies on the state and dynamics of electron transfer processes of exciplexes (a complex, existing in an excited state, that is dissociated in the ground state) has shown that micellar system alters the dynamics when compared to normal solutions. The exciplexes radical ions can be protected with success depending on expulsion from micelles directly after the initial electron transfer process. This micellar catalyzed ion-pair separation progression results in the formation of stable and long-lived ion radicals, which may well be detected both by transient absorption and by photoconduction methods. Alternatively, if the micellar surface "traps" the ion-pair owing to strong charge attraction, then a rapid ion recombination processes results. Such mechanisms are usually only observed in highly viscous systems but are readily observed in micelles and may help elucidate the catalytic activity of micellar systems [44]. Similarly, Alkaitis et al. studied the laser photolysis of phenothiazine in SDS micellar system compared to the photolysis in a methanol solution and found that the SDS leads to the formation of solvated electrons, cation radicals and triplets; therefore the yield of ions is much larger with an SDS system [45].

There are several possible routes for reactions to take in a micellar system. For a simple reaction of $A + B \rightarrow \{C\} \rightarrow D$, one must sometimes consider the transition state C and how it interacts in an aqueous micellar system. Both A and B could be inside the hydrophobic core of the micelle. There are three possible locations for A initially: partially in the micelle (Stern Layer, or Gouy–Chapman Layer or Palisade Layer), the surface of the micelle, or in the water. The same location options are true for B, D, and C. In fact, some reactions may be catalyzed by an anionic surfactant but quenched or retarded by a cationic one because of the charge attractions involved. One example is the free-radical reaction that is used for emulsion polymerization, a key industrial procedure that has a micellar system producing a proximity effect. The free radical that causes the polymerization and the reacting monomer are adjacent because of the micelle achieving an enhancement of the polymerization. Several other interesting systems are discussed by Thomas [7] and Fendler et al. [29] La Sorella presents a review in 2015 on the development of catalytic systems in water micellar systems [46].

Each of the characteristics of surfactants and micelles have not only been studied in their own right but have been put to use in many areas of chemistry, consumer products and in industry. Several reviews on the use of surfactants in analytical chemistry have been published previously [7, 19, 47, 48, 48, 50]. Not only does one expect the use of surfactants in water to replace organic solvents for greener chemistry, but also there is now an expectation that the micellar system will in many ways out preform its non-aqueous counterpart. Since the 1959 report of the catalyst of the phenol blue reaction with crystal violet in surfactant systems [40], there have been numerous investigations into the catalytic properties of micellar solutions and the mechanisms involved [51].

6.3 How environmentally safe are they? Concerns and beneficial uses

Green chemistry wishes to reduce or eliminate the generation of hazardous wastes. Surfactants replacing organic solvents can be a major aspect in this process by both replacement of a solvent and a reduction in the amount of chemicals used. However, one must continually try and look at updated literature on chemicals including their Safety Data Sheet to ascertain the viability of the current state of knowledge. Those of us who started our careers long enough ago remember the routine use of benzene, carbon tetrachloride and 2-naphthalamine in the laboratory. It took years of data and research to determine and understand the hazardous nature of these and other at one time common chemicals. Most anionic and nonionic surfactants are nontoxic and are widely used, and having a LD50 comparable to sodium chloride. Aquatic toxicity data are extensively available for the three major classes of surfactants: anionic, cationic and nonionic. It was found that the order of toxicity was cationic > anionic > nonionic surfactants [52].

Because of their widespread use in home and industry, surfactants commonly find their way into the environment [53]. The evaluation of environmental risk assessment and biodegradability of these organic substances is an important consideration for public health and environmental impact [54]. The toxicity data from laboratory and field studies are indispensable for us to evaluate the conceivable environmental risks from these useful chemical agents. The fate of the degradation of surfactants in the environment has been studied. They are readily degradable under aerobic conditions [55, 56] and also have been studied for biodegradability under anaerobic conditions [9]. Anaerobic conditions are found in the sludge digesters of wastewater treatment plants, sub-surface soil layers and the bottoms of rivers. After use, the surfactants are mainly expelled through a sewage treatment plant and then disseminated into the environment through the effluent released into surface waters and sludge disposal on agricultural lands. Surfactants have various behaviors and fates in the environment, which is exceptionally dependent upon whether their disposal is aerobic or anaerobic. Nonionic and cationic surfactants have been shown to have much greater sorption on soil and sediment than anionic surfactants [56]. An established initial step of wastewater treatment is the elimination of particulate matter in primary settling tanks; this particulate matter could include soap and detergents precipitated by calcium ions. Wastewater sludge is surfactant rich and is processed at elevated temperatures under anaerobic conditions [55]. Under these conditions, soap is readily biodegradable. Fatty alcohol sulfates, like sodium dodecylsulfate, and alcohol ether sulfates are readily biodegradable under both aerobic and anaerobic conditions [55].

6.3.1 Biodegradation

Biodegradation means the microbial breakdown of organic substances. The results of biodegradation for surfactants can be looked at on three levels. *Primary biodegradation* is where microorganisms cause the loss of surface-active properties that define the surfactant. *Ultimate biodegradation* is achieved when the surfactant is totally broken-down to inorganic end-products such as carbon dioxide, water and salts of any other elements, and consume these compounds and use them as energy and carbon sources. *Ready aerobic biodegradability* is an arbitrary classification of surfactants, involving certain specified screening tests for ultimate biodegradability. It is assumed that such surfactants in aquatic environment will then be able to rapidly and completely biodegrade under aerobic conditions.

The chemical structure of surfactants plays a major role in shaping their effect on the biotic and abiotic environment. This structure also influences the applicability of various analytical methods for analysis. Several factors must be considered: extraction and pre-concentration of the sample, qualitative and quantitative determination, and proper validation of the samples and method. It may occasionally be beneficial to separate anionic and nonionic surfactants simultaneously using solid phase extraction and to isolate them just prior to their quantitative analysis. A recent literature review provides an excellent starting point concerning the occurrence and concentrations of surfactants in different environmental samples [57]. The research summarizes the information on the analytical techniques and includes soil, street and indoor dust, bottom sediments, sewage sludge, and liquid samples, including precipitation, atmospheric deposits, aerosols, ground waters, surface waters, and sewage, along with their basic parameters of analysis, advantages and disadvantages of each method with numerous references [57]. Several classes of surfactants can be analyzed by potentiometric titrations including Epton's two-phase titration method [19]. The endpoint for a titration can be determined by turbidity and refractive index changes. In another type of titration, barium chloride, which forms a charged complex with nonionic surfactants result in a "pseudo-cationic" molecule, is detectable by a surfactant-sensitive electrode [19, 58].

6.3.2 Reclamation

In an opposite role for surfactants, they can be used as soil and water decontamination agents. Another admirable review emphasizes the currently surfactant-based soil and wastewater treatment technologies that clean the environment with green surfactant chemistry [59]. Hydrophobic pollutants (such as pesticides, petroleum hydrocarbons, PCBs and PAHs) are 100–1000 times more soluble in micelles than bulk water. The aromatic hydrocarbons are encapsulated in the micelles' hydrophobic interior and are removed through ultrafiltration since their size is larger than the pores of the filter [59]. Removal of heavy metals pollutants is possible by the micellar solution of surfactants [60, 61]. A flotation procedure using surfactants is normally used in mineral ore processing for separating hydrophobic and hydrophilic from one another, as in the purification of copper ore. This method can also be used as a remediation technique for polluted soils because it is effective in the removal of both organic and inorganic pollutants. The use of reverse micelles to remove ionic dyes like methyl orange and methylene blue was also discussed in this review paper [59].

Pharmaceuticals and personal care products have an increasing presence in the nation's waterways. Their removal is becoming a major problem of concern. Surfactant-enhanced extraction has been reported for the removal of different personal care products. One surfactants method is emulsion liquid membrane, which has acquired ample attention for the removal of these pollutants from water. The extraction method is based on liquid membrane technology for selective permeability of solutes with the micelles acting as both extraction and stripping agents. Sequestering of pharmaceuticals can be accomplished by using the enhanced adsorption effects that surfactants impart to a solution [59]. Shah et al. also review the various methods that surfactants can be employed for the removal of toxic metals by using ultrafiltration, CPEs, activated carbon (having 2–4 increased capacity with surfactants), soil washing/desorption/extraction, adsorption onto soil and phytoremediation. The use of biosurfactants in heavy metals removal is reported to be more effective than their synthetic counterparts. Biosurfactants are regarded as having lower toxicity and better biodegradability with improved stability over a wide range of temperature and pH conditions, ionic strength/salinity and exhibiting enhanced foaming properties [59].

6.3.3 Linear alkylbenzenesulfonates in the environment

Among the anionic surfactants, the linear alkylbenzenesulfonates (LAS) are one of the most popular laundry detergents with billions of pounds produced each year. Naturally their environmental fate has been well studied. The linear chains are much more biodegradable and less toxic than those with branched chains. The fate of surfactants has allowed chemist to look at these chemical from a variety of ways.

Besides their nature in forming micelles from their hydrophobic portion, they are ionic chemicals with electrostatic interactions. There are three types of interactions: hydrophobic interactions from the organic tail, chemical interactions of the sulfonate head group and electrostatic interactions from the charge on the sulfate group and the inorganic materials and humic material in a sediment. These options for interaction make the study of the exact mechanism of sorption more difficult because changing one parameter of an experiment actually changes the other types of interactions as well [62]. The solution's pH and ionic strength (types of ions too) also affect the behavior of the surfactant in solution and its sorption onto solid material. A change in pH can affect surfactant and surface charge of the material. For example, under very low pH conditions the sulfate group would theoretically lose its charge by being protonated, and the surfactant would behave as an entirely nonpolar alkane.

The formation of complexes between LAS and cationic surfactants such as alkyltrimethylammonium chloride and dialkyldimethylammonium chloride unfortunately results in the complex adsorption onto river sediments, giving biodegradations rates that were two to three times longer than LAS alone [55]. Kruger et al. reported the rates of biodegradation improved with increasing dissolved oxygen concentrations. Results showed that there was preference for the biodegradation of the longer alkyl chain LAS homologs and external isomers (i. e., 2- and 3-phenyl), but that laboratory results were two to three times greater than those experienced in field tests. Kruger suggests based on his results that a supplementary increase in the injected dissolved oxygen concentration during the continuous field test would have caused an amplified biodegradation rate [63]. In a review article, McAvoy et al. reported that concentrations of LAS in anaerobically digested sludge (10,462 ± 5170 μg/g) were one to two orders of magnitude greater than those detected for aerobically digested sludge (152 ± 119 μg/g), demonstrating that LAS is degraded more expeditiously under aerobic conditions [64]. Anaerobic conditions are found in the sludge digesters of wastewater treatment plants, sub-surface soil layers and the bottoms of rivers.

Others have reported similar results with escalated levels of biodegradation (97–99 %) having been establish in some water treatment plants with aerobic processes [55, 65, 66]. By comparison, alkyl phenol ethoxylates are less biodegradable with values of 0–20 % having been cited [55]. The comprehensive biodegradation of surfactants necessitates a consortium of bacteria due to the partial metabolic capacities of different microorganisms [55]. The risk assessment of this class of surfactants to terrestrial plants and animals was described by Mieure et al. who also determined that there are satisfactory margins of safety in the use of wastewater for the irrigation of crops [67].

Once sludge from a treatment plant is applied on land, the LAS are promptly metabolized by aerobic bacteria and do not accumulate in soil as demonstrated by field experiments that showed that the application method, and whether the soil had been ploughed, or not, had no effect on degradation rates of the LAS [56].

Anionic surfactants can be found in soils as a result of sludge application to crop land and wastewater irrigation of farms. High concentrations of surfactants together with the metals associated with them can embody an environmental risk. However, at low concentrations, surfactant application to crop land and soil is unlikely to have a significant effect on trace metal mobility [68, 69]. Edwards et al. reported on the distribution of nonionic surfactant Triton X-100 and phenanthrene in a sediment/aqueous system and found that it can act either to heighten or to impede phenanthrene sorption from bulk solution [70]. Triton X-100 is a alkylphenol ethoxylate surfactant, these may not be appropriate for field remediation work, owing to these types of surfactants degrade into undesirable alkylphenol monoethoxylates and diethoxylates in addition to alkylphenols during the course of anaerobic biodegradation [71]. It is biodegradable under aerobic conditions, however.

Soap and other surfactants will react with Mg^{2+} and Ca^{2+} to form solids. These tend to adsorb with solid particles. Only those dissolved in water can be metabolized by microorganisms. Cationic surfactants are absorbed onto the anionic sludge at the bottom of landfill sediment, water treatment plants, bottom of septic tanks, river bottoms and lakes deeper than 10 m [9]. Anaerobic biodegradation requires the co-operation of different types of microorganisms like those found in a complex food chain to accomplish complete biodegradation. Merrettig-Bruns and Jelen reported in detail on the anaerobic biodegradation of a number of surfactants in all four classes. They concluded that heterogeneous atoms like ester bonds in the chemical structure improve the anaerobic biodegradability of those surfactants significantly. Esterquats, unlike quaternary ammonium compounds, are also ultimately biodegradable under anaerobic conditions, and this is one of the reasons esterquats have replaced quaternary ammonium compounds in common usage [9].

6.3.4 Perfluorooctanesulfonate in the environment

One exception is perfluorooctanesulfonate, PFOA, surfactants labeled as a persistent organic pollutant with the US EPA, Environmental Protection Agency, which established health advisories at the 70 parts per trillion level in 2016 for PFOA and PFOS and which had earlier a voluntarily agreement with industry to stop PFOA production in 2006 [72]. Water treatment with activated carbon or reverse osmosis can be used to clean the water. The two popular surfactants: linear alkylbenzene sulfonates and the alkyl phenol ethoxylates, APE, have recently been considered to be less desirable for widespread use also. Besides their use as detergents, and cosmetics, they find industrial uses in paints, pesticides, textile and petroleum recovery chemicals, metal working. Since they break down in the aerobic conditions found in sewage treatment plants and in soil to the metabolite nonylphenol

(4-(2,4-dimethylheptan-3-yl)phenol), which is not readily biodegradable. The polyoxyethylene chain appears to be easily biodegradable, but the NP derivative looks more resilient [55]. Nonylphenol is alleged to be an endocrine disruptor owing to its capacity to mimic estrogen and subsequently to disrupt the natural balance of hormones (Figure 6.3) [73]. Prior to these studied APE's accounted for about 55 % use in industry. The foremost alkylphenols consumed are nonylphenol and octylphenol. Nonylphenol ethoxylates cover about 80 % of the global market, and octylphenol ethoxylates account for the remainder [74]. In 1991 when Ana Soto of Tufts Medical School detected that breast cancer cells, which generally multiply only in the presence of an estrogen, showed the same behavior in plastic containers, apprehensions about potential nonylphenol estrogenic activity developed. Investigative work revealed that nonylphenol caused the growth [74]. The distribution of nonylphenolic compounds was found in the digested sludge of the Swiss sewage-waste treatment plants with 95 % nonylphenol and 5 % short-chain ethoxylates, partially since there is a hydrophobic nature in nonylphenol and partially because the anaerobic digestion of the sludge created nonylphenol [74]. These and similar results have resulted in regulations in the EU. On September 2014 the EPA recommended a Significant New Use Rule to oblige an EPA review before a manufacturer starts or resumes use of 15 different nonylphenols and nonylphenol ethoxylates. They have also used the Safer Detergents Stewardship Initiative to encourage a voluntary phase out of these products in industrial laundry detergents. The EPA has set that nonylphenol concentration should not exceed 6.6 µg/L in fresh water and 1.7 µg/L in saltwater [73, 75]. They are still used in much lesser quantities in the laboratory.

The toxicity of nonionic surfactants is contingent upon their structure: with increasing alkyl chain length there is generally an increase toxicity, while increasing ethylene oxide groups will usually reduce toxicity. These trends are comprehensible when one contemplates the toxicity mechanism of surfactants, specifically membrane disruption and protein denaturation, which are a function of the surface-active properties of surfactants [19].

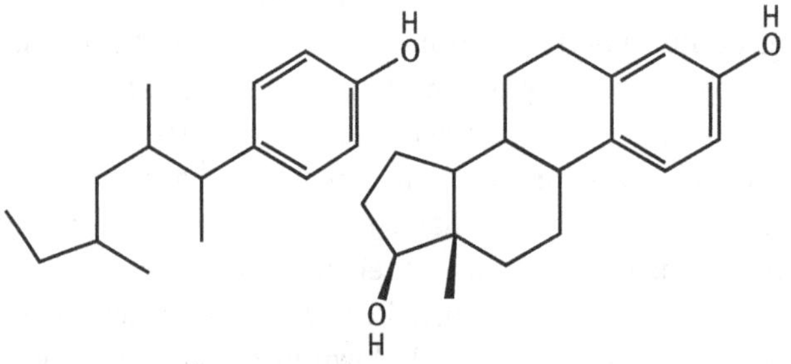

Figure 6.3: Structure of a nonylphenol (left) beside the estradiol hormone.

There has also been some concern about how the solubilization effects of micelles would have on other organic compounds being introduced into the environment, specifically polyaromatic hydrocarbons and pesticides. Aronstein et al. studied the effect of low concentrations of surfactants on the biodegradation of sorbed aromatic compounds (<0.01 %) in soil, because of the possible effectiveness of surfactants for stimulating the microbial destruction of pollutants. Alfonic 810–60 (a linear alcohol ethoxylate nonionic surfactant) and Novel II 1412–56 (a similar linear alcohol ethoxylate nonionic surfactant of slightly longer length) increased the extent of desorption of phenanthrene from a mineral soil. Both surfactants at 10 µg/g of soil noticeably improved the amount of biodegradation of phenanthrene in both the mineral and the organic soil. Biphenyl mineralization in the mineral soil was not altered by either surfactant, but biodegradation in the organic soil was improved by Alfonic 810–60. Surfactants at low concentrations may allow for the mineralization of sorbed aromatic compounds in polluted soils was one of their major conclusions, but not all the surfactants studied were successful [76]. In a study of the solubility of DDT and trichlorobenzene, as anticipated, the solubility was improved when the surfactant was present at concentrations above their CMC [56].

6.3.5 Cationic surfactants in the environment

The major uses of cationic surfactants are as fabric softeners and antiseptic agents against bacteria and fungi, cosmetics, in laundry detergents, mouth wash, used in synthesis of gold nanoparticles, and in industry [56]. The situation for cationic surfactants is more troublesome environmentally. Dialkyldimethylammonium chlorides have very low acceptable LD–50's, but alkylbenzyldimethylammonium chloride has an LD50 of 0.35 g/kg. Prolonged exposure of skin to surfactants can cause chafing as the surfactants disrupt the lipid coating that protects skin cells [77]. These surfactants are known to be toxic to animals, ecosystems, or humans and can also increase the diffusion of other environmental contaminants because of their unique properties [78]. Having a positive charge, cationic surfactants have a strong attraction for the surface of particulates in sewage sludge, which are principally negatively charged. Several studies have shown that in activated sludge 95 % of the cationic surfactants were adsorbed to the surface of particulate matter [55]. Investigations about alkylbenzyldimethylammonium chloride showed it to be ultimately biodegradable with >80 % of the carbon-14 labeled surfactants being released as $^{14}CO_2$ [55]. Under aerobic conditions, the biodegradability of quaternary ammonium cationic surfactants generally exhibit reductions with the number of non-methyl alkyl groups, and substitution with a benzyl group can lower the biodegradability even more [56].

In a review by Scott et al. the data available suggest that raw sewage passing through a modern waste treatment plant has a substantial quantity of its surfactant load eliminated. Aerobic treatment processes seem to deliver the best conditions for prompt primary and ultimate biodegradation via a variety of bacteria. Wastewater effluent released into the environment appears to have had its surfactant load reduced to the extent that lethality on aquatic organisms is slight. Superfluous safety margins exist for over 25 varieties of organisms [55]. The biodegradability of surfactants in the environment was summarized by Yang et al. fatty acid esters, and cationic surfactants although were judged to be persistent under anaerobic conditions but were found to be biodegradable in aerobic conditions [56]. The breakdown of the cationic surfactants in coastal waters was reported with an associated increase in bacterio-plankton density, signifying that the degradation occurs because the compound is used as a growth substrate [56].

6.4 Surfactants in analytical chemistry

There are several reviews of micelles in analytical chemistry [47, 50]. Although solubilization is the property most often thought of for micelles, because there are several sites available within its structure, one must also consider kinetics, and changes in the spectral profile that can also occur. Most of the UV visible absorption methods that have been reported involve the determination of metal ions through complexation with of chelometric indicators. This includes the determination of metals with PAN {1-(2-pyridylazo)-2-naphthol, CAS No. 85–85–8} using Triton X-100 as the nonionic surfactant for the determination of cobalt at 620 nm. The absorptivity was 1.9×10^4 over the range of 0.4–3.2 ppm [79]. For zinc using Triton X-100 at 555 nm with an absorptivity of 5.6×10^4 over the range of 0–100 ppm [80]. With TAN, 1-(2-thiazolylazo)-2-naphthol, using Triton X-100 for the determination of nickel at 595 nm after a 5-min reaction time at a pH of 9.2 with an absorptivity of 4.0×10^4 over the range of 11–110 ppm [81]. Also TAM was used with Triton X-100 for the determination of nickel at 560 nm with an absorptivity of 6.5×10^4 over the range of 0.12–1.20 ppm in soil samples [82].

The solubilization property of micellar systems alleviates the need for organic or mixed aqueous–organic solvents and is indeed its major advantage in analysis. This solubilization ability can alone be utilized for sample preservation and storage. One example shows that Brij-35 surfactant in the sample solution was as effective as a 40% acetonitrile solution in inhibiting the loss of polycyclic aromatic compounds' adsorption on the surface of borosilicate glass, or other containers [20].

6.4.1 UV–Visible spectroscopy

The use of surfactants to replace organic solvents in UV–Visible spectroscopy has been one of the major areas of success for this type of green analytical chemistry. There is often a 10-fold increase in sensitivity (molar absorptivity), as well as, a bathochromic shift reported with a change in solvents due to differences in solvent polarity. An early review by Hinze highlights many of the methods that had already been reported by 1979 [83]. It seems plausible that once the electrostatic forces have brought together the oppositely-charged molecules, hydrophobic interactions occur, dramatically altering the micro-environment experienced by the chromophore or complex. Nonionic surfactants exhibit a behavior that is similar to that of organic molecules, without the electrostatic nature seen with cationic and anionic surfactants that can parallel the behavior of electrolytes. A good example of this is the reaction of cobalt with thiocyanate which in water forms the pink complex $[Co(H_2O)_6]^{2+}$ but in an alcohol or acetone mixture with water forms the familiar $[Co(SCN)_4]^{2-}$ blue complex that absorbs at 625 nm. This was the basis for the visible detection of cobalt using a 3 % Tween-80 nonionic surfactant that eliminates the use of acetone, isoamyl alcohol or other organic solvents in a flow injection analysis method [84]. In many cases there have been reports that indicate that there is a significant bathochromic shift in the absorption wavelength for metal chelate complexes in the presence of micelles. These shifts were used to determine the CMC of some surfactants and the changes in an indicator's pKa have also been widely studied [47]. Evidence supported by UV–Visible and fluorescence studies has caused speculation that a surfactant molecule can complex with the metal complex changing its character. Sometimes call ternary complexes, they have different spectral properties from the binary complex that is usually formed between the metal and the ligands. Rather than forming a separate ligand the "complexation" may occur at the surface of the micelle's charged layer as was shown to happen with the π electrons of pyrene.

6.4.2 Cationic surfactants

A large number of metal cations have been analyzed with triphenylmethane dyes using the cationic surfactants CPC, CPB, CTAB, CTAC and tetradecyldimethylbenzyl-ammonium chloride, zephiramine {CAS No. 139–08–2} including F, Ti, Be, Ga, V, Sc, Y and Al. Also in this group were a number of lanthanides using pyrocatechol violet {CAS No. 115–41–3} and a cationic surfactant media [49]. In some cases the surfactant only seems to be increasing the solubility, while in others it may also be altering the pKa of the dye and/or actually actively participating as a chelating agent in the metal–dye complexation. In a study by Marczenko and Jarosz the experimental conditions for the formation of ternary complexes of aluminum cation with Eriochrome Cyanine R, Chrome Azurol S or Pyrocatechol Violet, in the presence of the surfactants zephiramine, CTAB or CPC formed a tertiary complex with the metal that exhibited greater molar absorptivities than the binary system. Complexes with Pyrocatechol Violet were reported as not suitable for a spectrophotometric method of analysis, however [85].

Exactly how the surfactant interacts was the subject of a contemporary study in 2016 on the binding characteristics between Alizarin Red S {CAS No. 130–22–3} and cationic surfactants [86]. Another series organic dyes that were studied were the Alizarin green dyes used to determine a variety of cations including: vanadium, indium and uranyl [49, 87, 88]. The reaction for nitrite analysis which was determined by the coupling reaction of *p*-nitroaniline with 8-hydroxyquinaldine {CAS No. 826–81–3} to produce the purple azoxine dye if carried out in CTAB did not require an organic extraction, and had $\varepsilon = 4.72 \times 10^4$ L/mol. cm which is 22 % larger than the earlier reported extraction method [89, 90]. There have been cases reported where the anion of the cationic surfactant used affects the analytical results of an experiment. This phenomenon is attributed to the reaction taking place or the product being physically located on the cationic surface of the micelle and the ion exchange properties of the anion in the solution being the driving force behind the observed data. The order of sensitivity was reported to be $SO_4^{2-} < Cl^- < Br^- \leq NO_3^-$ [49]. When 3,3-dimethyl-2-phenyl-3H-indol {3,3-dimethyl-2-phenylindole, CAS No. 6636–32–4} was used in a study by Sarpal et al. as a fluorescence probe studying to pKa's in SDS, CTAB and water, it was shown that CTAB offered a more hydrophobic environment for the probe molecule [91]. Cationic micellar systems generally enhance the acid dissociation and therefore decrease the pKa of organic compounds, and as expected an anionic surfactant will cause an increase in the pKa.

Cationic surfactants are known from catalysis studies to facilitate nucleophilic substitution reactions. There are several early examples of using this to enhance visible spectroscopy determinations. The reaction of cyanide ion with 5,5′-dithiobis (2-nitrobenzoic acid) to displace the corresponding absorbing thiol anion, using CTAB decreased the reaction time from 25 min to 1–3 min [92]. Micellar catalysis of nucleophilic reactions was used for determination of aromatic aldehydes, amines, and oximes with UV–Visible spectroscopy by catalytic acylation with *p*-nitrophenyl acetate in the presence of CTAB [93]. In 1979 Conners et al. reported the spectrophotometric determination of amino acids and peptides after CTAB-catalyzed reaction with 1-fluoro-2,4-dinitrobenzene. The more hydrophilic amino acid reactants exhibited larger relative rate enhancements, as projected for micellar reaction catalysis [94]. Recently the use of CTAB and Alizarin green were described in the analysis of benzalkonium bromide, used in eye drops, by the decrease in absorbance recorded in alkaline solutions [95].

6.4.3 Nonionic surfactants

The use of nonionic surfactant Triton X-100 has been used with the determination of metal complexes that normally use organic solvents like PAN, 1-(2-pyridylazo)-2-naphthol {CAS No. 85–85–8} and aqueous soluble reagents like PAR, 4-(2-pyridylazo) resorcinol {CAS No. 1141–59–9} [49]. The chromophore

Cadion, 1-(4-nitrophenyl)-3-(4-phenylazophenyl)triazene, {Cas No. 5392–67–6} was used with *p*-nitrobenzenediazoaminobenzene-*p*-azobenzene in Triton X-100 for the dual wavelength determination of cadmium [96]. The naphthalene analog of Cadion, Cadion 2B, N-[(4-nitronaphthalen-1-yl)diazenyl]-4-phenyldiazenylaniline {CAS No. 6708–61–8} was used for the determination of silver in a Triton X-100 micellar system [97]. Cadion2B was also used to determine cyanide by the suppression of the absorbance of the copper and silver complexes by that anion in a Triton X-100 surfactant system [49].

6.4.4 Anionic surfactants

There have been hundreds of reports of metal ion complexing agents (chelometric indicators) being analyzed in cationic and nonionic surfactant system but few that have used anionic surfactants [47]. A spectrophotometric determination for uranium(VI) was based on formation of a red–violet complex from the reaction with 2-(3,5-dibromo-2-pyridylazo)-5-diethylaminophenol {CAS No. 14337–53–2} in a SDS micellar system [98]. The same chromophore was used for the determination of zirconium in aluminum and steel alloys with SDS at pH 4.6 [99] and silver [100] and zinc [101] in an SDS system. SDS was also used as the solubilization system for the spectrophotometric determination of Co (II), Ni (II), Cu (II), Pd (II), Ru (III) and Mo (VI) using sodium isoamylxanthate {CAS No.: 2540–36–5} as a reagent [102]. Ghaedi reported the interference-free spectrophotometric determination of Ni (II) ions based on the reaction between nickel and α-benzyl dioxime (*N*-[(*E*)-2-nitroso-1,2- diphenylethenyl]hydroxylamine) {CAS No. 23873–81–6} in a SDS micellar in 2007 [103]. The non-aqueous use of SDS (15 % in acetone) was used to analyze the fluoride ion concentration in bottled and sea water using the fluoride/lanthanum (III)/Alizarin fluorine blue {CAS No. 3952-78-1} ternary complex [104]. A new spectrophotometric method for hemoglobin analysis at 534 nm using SDS was reported, which unlike other methods avoids oxidative reagents and does not produce toxic wastes such as KCN and NaN_3 that can cause environmental pollution [105]. The spectrophotometric determination of germanium with phenylfluorone, (2,6,7-trihydroxy-9-phenylxanthen-3-one) {CAS No. 975–17–7} at trace concentrations in a SDS surfactant system at 504 nm was reported by Dagar et al. [106]

6.4.5 Fluorescence spectroscopy

There are several reviews of fluorescence in micellar media [107, 108]. The Wandruszka review from 1992 covers the specific changes that occur in micellar media like Krafft point, changes in microviscosities and the use of fluorescence reagents to determine CMC and other attributes of micellar structure and would serve as a good introduction to the topics covered [17]. Although there had been many reports of fluorescence enhancement as part of a physical chemistry, probe characterization or catalysis study, many of these did not include any quantitative analysis of the analyte. The review by Hinze et al. in the 2008 *Encyclopedia of Analytical Chemistry* covers some of the same material but has a greater number of references (876) and the scope is more toward analytical applications and studies [20]. This reference includes several tables of applications: (1) drugs, vitamins, dyes and other organic substances, (2) determination of organic analytes, (3) determination of selected inorganic species and (4) micellar-enhanced lanthanide-sensitized determination of organic species.

In 1972 Ishibashi reported a six-fold enhancement in the determination of aluminum using lumogallion {CAS No. 4386–25–8} as the chelating agent combined with the nonionic surfactant IGEPAL CO 890 {polyoxyethylene (40) nonylphenyl ether, CAS No. 68412–54–4} [109]. San–Medel et al. studied the lumogallion niobium fluorescence with nonionic surfactant Triton X-100 and concluded that only when CMC is established can the ternary complex be accommodated in the surface of the micelles and fluorescence enhancement is observed. They concluded with the over-all idea of maximum fluorescence enhancements is detected when electrostatic and hydrophobic interactions can act concurrently for the complex and the micelles. They collated their results with the structure of the complex and that of the surfactants [110]. Some reports indicate that part of the fluorescence enhancement observed in micellar systems is not just due to increased solubility, encapsulation and protection of the fluorescence species from collisional deactivation inside the micelle, or interaction of the excited state with an ion-pair of the surfactant, but also the solubilization of the quenching impurities bound inside a separate micelle so that they cannot come in contact with the analyte which is in another micelle. The primary factor accountable for enhanced lifetimes in micellar media seems to be reduced quenching constants [20]. In 1987 San–Medel et al. investigated and reviewed several classes of metal chelating agents (flavonols, 8-hydroxyquinoline derivatives, azo dyes, and anthracene derivatives) for the solubilization in micellar media had fluorescence with the metal cations of Al, Nb and Ta. They observed that greater enhancement was observed when not only did the surfactant solubilize the complex at the CMC but when there were electrostatic interactions that occurred between the surfactant and the complex resulting in a more rigid structure. For some systems that exhibited fluorescence quenching in the presence of the cationic surfactant, they showed that this was due to inter-systems crossing being enhanced and phosphorescence occurring at the expense of the fluorescence [111]. Dominguez et al. also focused on the niobium–lumogallion–tartrate system in a 1989 paper that discussed the competition between the ligand's and the metal complex's interaction with the micelles for a variety of nonionic surfactants and CTAB where there was a noted red shift in the excitation wavelength and a blue shift in the emission wavelength from 630 nm in water to 600 nm in Triton X-100 [112].

The early work in 1982 of Singh and Hinze looked at the intensity of the pyrene fluorescence and its enhancement from 3 to 16 times in micellar systems of CTAC, SDS and Triton X-100 when compared to ethanol. The spectral parameters, fluorescence lifetimes, quantum yields, lower detection limits, and analytical figures of merit for pyrene all four systems were compared [113]. In a separate paper they investigated the 8- to 20-fold enhancement effects of different surfactant micellar systems upon the spectrofluorimetric method for the determination of amino acids by Roth's method and the dansyl chloride procedure. They found that the fluorescence intensity of dansyl glycine was enhanced when in the presence of CTAC or dodecyl(trimethyl) azanium chloride {CAS No. 112–00–5} and the zwitterionic surfactant *N*-dodecyl-*N,N*-dimethylammonium-3-propane-1-sulfonic acid {CAS No. 14933–08–5} micellar systems. Similarly, the lysine derivative of *o*-phthaladehyde-2-mercaptoethanol exhibited increased fluorescence in the nonionic Brij-35 or Triton X-100 and SDS surfactants [114].

A study of the fluorescence of the Pb-morin system in the presence of 2% the nonionic surfactant Genapol PF–20 (a poloxamer which is a nonionic triblock copolymers composed of a central hydrophobic chain of polyoxypropylene bordered by two hydrophilic chains of polyoxyethylene) was enhanced about 9-fold, at a pH of 3.3. The presence of a nonionic surfactant also gives greater stability with the system being stable for at least 3 hours. The excitation maximum occurs at 420 nm with fluorescence occurring at 495 nm and the detection limit was reported to be 0.06 µg mL^{-1} [115]. A 10-fold increase in the determination of aluminum with morin in the presence of a similar surfactant system was reported at a pH of 3.8 and a detection limit of 0.2 ppb. The number of interfering ions was also reduced significantly [49]. The pH can be a major factor in the enhancement and the type of surfactant that is best for the analysis. An anion that is the excitable molecular form would be expected to have the highest enhancement with a cationic or nonionic surfactant rather than an anionic surfactant which might produce a quenching effect. Fluorescent complexes of 8-hydroxy-7-iodoquinoline-5-sulfonic acid {CAS No. 547–91–1} with Al, Mg, and Zn cations in cationic surfactant systems exhibited enhancements in CTAB [20]. For the fluorescence analysis of Ga(III) with 1-(2-pyridylazo)-2-naphthol, PAN, {CAS No. 85–85–8} the anionic SDS micellar system increased sensitivity about 20 times better than that achieved in a 20:80 (v/v) ethanol-aqueous solvent system [20].

The fluorescence intensities of terbium, europium and samarium complexes with several β-diketone derivatives in the absence and presence of tri-*n*-octylphosphine oxide (TOPO) in micellar solution of nona-oxyethylene dodecyl ether [116]. Various analytical applications, including immunoassays, quantification of organic compounds has utilized lanthanide and actinide ions complexed with organic ligands for the sensitization of fluorescence analysis [20]. The presence of another f–block element can cause the fluorescence to become even more sensitive. Fluorescence intensities of Eu^{3+}, Sm^{3+}, Dy^{3+} and Tb^{3+} in the presence of a surplus of cations of Y, Lu, Gd or La, chelated with pivaloytrifluoroacetone in a Triton X-100-ethanol solution containing 1,10-phenanthroline, were increased by factors ranging from 61 to 1078 fold [117]. This can be referred to as the cofluorescence effect and has been reviewed by Xu et al. [118]

Fenproporex {3-(1-phenylpropan-2-ylamino)propanenitrile, CAS No. 16397-28-7} produces amphetamine as a metabolite and has been used as an appetite suppressant was enhanced by a factor of 2.6 when analyzed in the anionic surfactant SDS [119]. A similar study was conducted on benzodiazepines (a class of psychoactive drugs) using surfactants provided micellar enhancement factors for their fluorimetric analysis in the range 1.2–6.5 increase, depending on the nature of both the benzodiazepine and the surfactant used [120]. Tetracyclines form luminescence complexes with Eu^{3+} and it was established that the emission was enhanced by surfactants. Both CPC and Triton X-100 surfactants were studied, and enhancement by a factor of up to 34 was detected for some tetracyclines. Small structural variances between the different tetracyclines examined had a noticeable effect on the intensity of emission [121]. The analysis of EDTA, ethylenediaminetetraacetic acid, in various foods was accomplished with the fluorescence ternary complex Zr(IV), EDTA and Alizarin Red S in CTAB with a detection limit of 3.4 ng/mL [122].

6.4.6 Phosphorescence spectroscopy

The use of phosphorescence analysis has been limited despite the low detection limits possible because of the cryogenic temperatures that had previously been necessary and the numerous possibilities of quenching of the reaction. The unique feature of phosphorescence is the time delay due the triplet states which have long lifetimes. This was used to study probes of a micellar system from nanoseconds to seconds. Some early investigations on triplet anthracene indicated that the exit time of this molecule from a CTAB micelle was of the order of one millisecond [36].

Oxygen quenching of the triplet state has been a major difficulty in phosphorescence. It has a much lower solubility in water than in organic solvents, and one would expect a higher oxygen concentration in a micelle core than in the aqueous phase, a condition which is found experimentally. At pressures below 1 atm of oxygen the ratio of micelles with oxygen is about 1 in 7. Nevertheless, dissolved oxygen moves freely in and out of micelles, which results in the quenching of micelle–bound excited states [7]. Sanz–Medel et al. describe using sodium sulfite as an oxygen scavenger in surfactant systems for micelle–stabilized room temperature phosphorescence [123]. Other methods of oxygen removal were summarized by Hinze in his review [20].

Micelle-stabilized room temperature phosphorescence (MS-RTP) was first reported by Kalyanasundaram et al. [124] Some of the early reports of room temperature phosphorescence (RTP) utilized SDS with heavy atom cations of either silver or thallium (which is toxic) to increase the quantum yields in the determination of a number of polyaromatic hydrocarbons [125], and later of carbazole and its derivatives [126]. Because the micelle concentrates the interacting species in a far smaller volume about the micelle that has a heavy atom as its counter-ion, this effectively intensifies the heavy-atom concentration around the analyte and, the subsequent phosphorescence [48]. A series of licit and illicit drugs were investigated with RTP, sensitized phosphorescence and fluorescence using micellar media and cyclodextrens [127]. Thiabendazole (2-(4-thiazoly)-1H-benzimidazole), a fungicide widely used in agriculture, was determined in pineapple by RTP in a SDS micellar system using thallium as the heavy atom [128]. The investigation of 8-hydroxyquinollne and some of its derivatives as potential complexing reagents for the room temperature phosphorescence using CTAB for the determination of niobium with bromoform as the heavy atom source after oxygen removal was reported by Kalyanasundaram et al. in 1987 [129]. They concluded that electrostatic and hydrophobic forces acting simultaneously seem essential to secure micelle-stabilized room temperature phosphorescence. Similarly, Liu et al. used room temperature phosphorescence to determine a gallium(III)-7-iodo-8-hydroxyquinoline-5-sulfonic acid complex {Ferron} in the presence of CTAB and with sodium sulfite as an oxygen scavenger [130].

6.4.7 Atomic spectroscopy

The use of organic solvents in flame atomic absorption spectroscopy, FAAS, and emission methods of analysis has been less studied. To decide if an engine needs replacement bearings and rings it is a common procedure to use of organic solvents for the analysis of metals in engine oil to determine the wear characteristics of the engine. It is known that in the spray chamber organic solvents produce an aerosol that has a greater number of small droplets and the corresponding signals are usually increased compared to aqueous systems. However, there have been mixed results reported in the literature for surfactants as enhancement agents in FAAS. The anionic surfactant SDS exhibited enhancements in the FAAS absorption value while cationic and nonionic surfactants cause a depression or no effect in the signal [131]. Enhancement with SDS was also studied for several metals (Cr, Cu, Ga, In, Ge, Si, Sn, Te, Sb, As, Bi) and their effect on three electrode argon DC plasma spectrophotometer. The SDS enhancement effect was explained by a combination of earlier observed phenomena: effects of easily ionized elements, increased Penning ionization, thermal pinch and increased residence time in the plasma [132].

More consistent applications have been reported when the surfactants have been used for emulsifying agents to solubilize water-immiscible samples and avoid the use of organic solvents. This method has been used to determine lead in gasoline and lubricating oils, zinc in antifungal preparations and iron in lubricating oils [49]. surfactants have been used successfully to sequester metal cations for their later determination by FAAS using CPEs and some other preconcentration steps discussed later in the flow injection analysis section of this survey.

6.4.8 Chromatography

In 1979 Armstrong first reported the use of surfactants in the mobile phase for TLC analysis for polynuclear aromatics [133] and pesticides [134]. This was followed with their use in HPLC [135, 136]. The retention times were dependent upon the concentration of the surfactant and if the analyte reacted with them. Another factor that could contribute to the selectivity of the micellar system if the solute was ionic or not, and if it was it could form an ion-pair with a surfactant molecule [137]. The elution order was reversed when one went from the anionic surfactant SDS to the cationic surfactant CTAB in the separation of phenol and acetophenone [50]. The added advantage that micellar systems have with HPLC is not only the separation characteristics but also the enhanced detection of surfactant systems that has already been documented using an UV–Visible or fluorescence HPLC detector [48]. Addition of small amounts of other organic solvents, like *n*-propanol, have been reported to have great effects on the separation characteristics of the chromatographic separation and using a column at 40 °C, and helps overcome some earlier problems with efficiency [50]. Changing the concentration of the surfactant in the mobile phase during a separation mimics gradient elution of a secondary solvent in HPLC. A review by Hinze in 1989 [138], and Esteve–Romero et al. in a 2016 review article looked at micellar liquid chromatography being used for the analysis of several drugs in serum and urine including: anticonvulsants, antiarrhythmics, tricyclic antidepressants, selective serotonin reuptake inhibitors, analgesics and bronchodilators [139]. Micellar chromatography results in the production of less toxic solvents and lower cost of reagents, and therefore should be of considerable use in the development of a green chemistry HPLC method of analysis.

6.4.9 Micellar electrokinetic chromatography

Terabe et al. first reported the use of micellar systems in 1984 in the electrokinetic separation of 14 different phenols using SDS allowing this technique's samples to include water insoluble species. The theoretical plate count was between 210,000 to 400,000 for the 19-minute separation. The SDS was carried from the negative electrode to the positive electrode. When the cationic surfactant CTAB was used the electroosmotic flow was in the opposite direction [140]. The use of MEKC has been reported in varied field including pharmaceutical, clinical, environmental, and biochemistry for both organic and inorganic compounds. Its numerous advantages and usages were reviewed by Khaledi in 1997 [141], in 2012 by Sepaniak et al. [142], in toxicology [143], element speciation [144] and in the analysis of pharmaceuticals

[145]. The review by Silva incorporates a review of instrumentation and analytical methodology including the use of micellar systems for online sample concentration techniques and detection [146].

6.4.10 Electrochemistry

In classical polarography surfactants were used to eliminate polarographic maxima. Since that early use there have been numerous physical chemistry studies using electrochemistry to help understand the fundamental nature of micelles. Although surfactant systems can fulfill their normal role of increasing solubility, they can also affect the characteristics of the double layer at the electrode and the diffusion of active species through a solution or even act as masking agents for some species [50]. The review by Rusling gives not only a historical review of surfactants in electrochemistry but also the background of these organized media [147]. Specific examples include the effect that adsorbed surfactants have on electrode behavior, including the lowering of the differential capacitance and reorganizing of the surfactant on the surface of the electrode due to changes in potential. Results discussed show evidence of adsorbed bilayers or hemimicelles on Pt and Hg electrodes [148]. The adsorbed surfactant can alter the double-layer arrangement, and the rate of electron transfer (by both acceleration and inhibition), and the apparent half-wave potential of an electroactive analyte [48]. In 1952 Proske used the solubilization power of micelle media to report the polarographic determination of anthraquinone in water using the surfactant Aerosol MA, dihexyl sodium sulfosuccinate {CAS No. 3006–15–3} [146]. Besides the solubilization ability of surfactants and the reaction intermediates, the diffusion process to the electrode is altered by their presence. Experiments produce a variety of results depending on the surfactant and the analyte. Like the alteration of the pKa in acid/base indicators the microenvironment of the micelle may cause a shift in the reaction potential. Generally, the half-wave potential has a negative shift but for some analytes there was no change and for the analysis of tetrathiofulvane, 2,2′-bis(1,3-dithiolylidene), in CTAB there was a positive shift [49]. In cyclic voltammetry studies the increased solubility of micellar systems can evince reversible reactions not seen in normal solvents. Micelles can stabilize anion radicals formed in the course of reduction reactions for some organic compounds. Such was the case for phthalonitrile and fluorenone and this behavior was attributed to the increased solubility of the anion radical in cationic micellar media [49] and with studies of nitrobenzene [149]. The micellar media has also been used because it allows for the solubilization of the electrochemically generated titrant to allow the electron transfer as discussed in the case of the ferrocinium ion being generated from ferrocene {Cas No. 102-54-5} [49].

In a more recent literature survey from 2006 by Vittal et al. which includes the modification of electrodes with metal hexacyanoferrates, their derivatized oxides and titanium oxide are reviewed [150] including the work they first reported of CTAB and nickel hexacyanoferrate films characterized by cyclic voltammetry [151].

6.4.11 Extractions

Surfactants have been used to form close ion-pairs that are soluble in organic liquids to achieve the separation of metal ions from aqueous solutions. Often this is then followed by the spectrophotometric determination of the species. Some metals form anionic complexes with other chelating agents so cationic surfactants may be used to form close ion-pairs with these species and extracted (in carbon tetrachloride, chloroform, xylene, or dichloroethane) and analyzed [49]. Other ions have also been used to form the ion-pair and then that is soluble in a nonionic surfactant [152]. This method has also been used for the determination of surfactants in water. Because many of these procedures use organic solvents we shall bypass a discussion of them but Hinze reviewed some of these in 1979 [83] and Paleologos in 2005 [153]. Reviews on extraction and pre-concentration of organic pollutants in environmental samples by Ferrera in 2004 [154] and extraction separation and preconcentration in metal analysis by Stalikas in 2002 are comprehensive guides to this area of analysis [155].

An alternative green extraction method uses the cloud point of the surfactant itself to effect the separation, known as CPE [22, 23, 156]. Upon changes in temperature, aqueous solutions of several nonionic and some zwitterionic surfactants develop a turbidity and a separation into two phases. This cloud point temperature occurs over a narrow temperature range where the large number of surfactants cause a scattering of light and make the solution appear cloudy. The process requires several easy steps. Add surfactant to solution to be analyzed making such its final concentration will be greater than its CMC value. After the species has dissolved raise the temperature in a temperature bath above the cloud point (or lowered in the instance of zwitterionic surfactants) and then phase separation occurs with nonionic surfactants in the lower layer or zwitterionic surfactants in the upper layer. The separation can be completed by gravity or centrifuging the same where the analyte has now been concentrated into the micellar layer. Some surfactants appear on the top layer and others on the bottom layer. Adjustment of the viscous surfactant rich layer may be desired before final analysis of the product [23]. The temperature at which a CPE happens can be changed by the addition of salts or other surfactants which may also alter the CMC [152]. The addition of sulfate increases the cloud point while urea and iodide lower it [22]. Extraction of metal ions requires smaller volumes and greener chemistry compared to traditional extractions. Watanabe pioneered this work and has written a review on the subject [157]. Extracted metal ions were then determined by a variety of analytical methods including fluorescence, visible and flame atomic absorption spectroscopy and others. Using the CPE technique as a sample preparation and preconcentration prior to analysis by HPLC or gas chromatography and an electron capture or flame ionization detection have been effectively used for the quantification of environmentally important compounds. This method produces fewer wastes and is less expensive than traditional extraction methods and more efficient is several incidences for a variety of organic and environmental samples [23]. The vast quantity of publications involving CPE involves the purification of membrane proteins following the work of Bordier [158]. A review by Quina and Hinze has several useful tables: (1) Some Extraction Parameters, (2) Recent CPE of Biomaterials/Clinical Analytes, and (3) Extraction and/or Preconcentration of Organic and Environmental Compounds (pesticides, fungicides, PCB's, PAH's phenols and others) and numerous applications in their article [23].

By combining CPE with fluorescence analysis one can achieve low detection limits. Using the nonionic surfactant Genapol X 80 {CAS No. 9043-30-5} the pesticides napropamide and thiabendazole were extracted and analyzed from water and soil samples. The fluorescence enhancement factors attained in the surfactant-rich phase were 1.2–2.0 for these two pesticides. Detection limits less than 0.2 µg/L and recovery rates of equal to 95 % were realized [20].

6.4.12 Titrations

There have been numerous investigations that use the ability of the surfactant to solubilize organic compounds that normally would have to be titrated in a non-aqueous media that can be titrated in a micellar environment in water using either a potentiometric or color indicator. This offers one the opportunity to use a greener solvent than the non-aqueous solvents normally used. Some examples of solvent substitutions are given in Table 6.1. As early as 1934 Hartley reported on the effect of surfactants on acid–base indicators studying some cationic surfactants such as CTAB [39]. The significant shifts in the apparent pKa values were also later reported to occur in nonionic micelle systems as well [49, 159]. For example methyl red has a pKa of 4.95 in water but the apparent pKa in a

Table 6.1: Comparison of solvents used in non-aqueous titrations and micellar systems.

Types of compounds titrated in micellar systems [49]	**Non-aqueous solvent replaced by micellar system [161]**
Long chain amines	Acetic acid, dioxane, acetonitrile and others
Fatty acids	Acetone, methanol/benzene, DMF
Benzoic acids	CS_2, acetonitrile, methanol
Sulfonamides	DMF
Sparingly soluble weak acids	Acetone, methanol/benzene, DMF
Sparingly soluble weak bases	Acetone, methanol/benzene, DMF
Halophenols	Pyridine, DMF, ethylene diamine, and others

micellar system changes with the type of surfactant to 3.67 (cationic), 6.63 (anionic), and 5.20 (nonionic) [50]. Examination of the data and the low dielectric constants reported support the simple electrostatic theory for this phenomenon [159]. Fernandez et al. used two different fluorescent pH indicators, a hydroxycoumarin and an aminocoumarin dye, that were incorporated by means of long paraffinic chain substituents to neutral, anionic, and cationic micelles and comparisons were made with the pKa's of aqueous solutions to determine if the electrical potential and/or a change of polarity were responsible for the apparent pKa shifts [160]. In their findings they concluded that the indicator's physical location and behavior were similar in SDS and Triton X-100, but different for CTAB. The electrical potentials for SDS micelles (−134 mV) and CTAB (+148 mV) have a comparable shift to the potentials measured by the lipoid-pH indicator in charged monolayers of analogous charge density.

Underwood observed for the titration of carboxylic acids in CTAB with NaOH, phenolphthalein indicator, and amines in SDS with HCl, phenol red indicator, that the electrodes gave sluggish results compared to visual indicators [162]. Staroscik reported using both visual and potentiometric endpoint detection for the determination of some barbiturates in cationic micellar systems by acid-base titrimetry [163]. A study was made by Gerakis et al. using microcomputer-controlled titrations of pharmaceutical compounds: aspirin, naproxen, iopanoic acid {(*RS*)-2-[(3-Amino-2,4,6-triiodophenyl)methyl]butanoic acid, CAS No. 96-83-3} and benzoic acid and its derivatives using Tween-80, CTAB, CPC and SDS. Direct titrations in CTAB had good correlations with the non-aqueous titrations counterparts for the weak acids [164].

6.4.13 Flow injection analysis

Flow Injection Analysis, FIA, was first reported by Ruzicka and Hansen [165]. This method of analysis combines the speed of chromatography with numerous detection methods. In it a sample is injected into a stream of solvent that includes the chromophore and they are mixed in a reaction coil before passing through the appropriate detector. The benefits of FIA include easy variation of the parameters, low solvent and chromophore concentrations, conservation of reagents, and high sample throughput. The method has been used for automation of many different analytical determinations, there have been periodic reviews in the literature [165–168].

A typical paper is the one by Resing and measures on fluorometric determination of aluminum in seawater by flow injection analysis with in-line preconcentration, using a resin, which employs the already existing lumogallion reaction in Brij-35 for the analysis [169]. The flow injection analysis of benzoyl peroxide in acne cream and flour using *N,N,N,N*-tetramethyl-*p*-phenylenediamine, Wurster's reagent {CAS

No. 637-01-4}, and a SDS surfactant system including cerium (IV) as a catalyst was measured by visible spectroscopy at 612 nm is another typical example of the use of micellar systems to act as a solvent and a catalyst for reactions. Other methods for benzoyl peroxide had heating baths and waiting times of 20–40 min [170].

The surfactant Triton X-100 illustrated the catalytic abilities of micelles in the FIA analysis of cobalt (II) and nickel (II) using PAR, {4-(2-pyridylazo) rescorcinol, CAS No. 13311–52–9}, in which the colored metal complex was formed without the need of a 15 min hot water bath. Detection was with a visible spectrophotometer at 510 nm [171]. Catalytic kinetic FIA spectrophotometric method for the determination of nanogram amounts of copper (II) was created using the catalytic effect of copper on the reduction of azure B, {3-Methylamino-7-dimethyl-aminophenothiazin-5-ium chloride, CAS No. 531–55–5} by sulfide in a CTAB surfactant system at 647 nm, with 50–1600 µg/L as the calibration range and 9.2 µg/L as the detection limit [172].

Zinc in seawater was analyzed by a FIA spectrophotometric method at 630 nm in a CTAB borax buffered solution using salicyl-fluorone {2,6,7-trihydroxy-9-(*o*-hydroxyphenyl)-3-fluorone, CAS No. 3569–82–2} as the metal complexation agent with a detection limit of 1.5 µg/L [173]. Lead determination was reported in the range of 1.0–12.0 µg/L and a detection limit of 0.027 µg/L employing 1,5-diphenylthiocarbazone {CAS No. 60–10–6} as the complexing agent in sulfuric acid and SDS for a FIA spectrophotometric analysis at 500 nm [174].

A very interesting use of surfactants is reported in FIA methods have been developed that will accomplish extractions and preconcentration of an analyte. The use of a complexation agent immobilized on a surfactant-coated alumina column acts as a separation and preconcentration step has been the basis for an analysis by Ahmadi et al., which involves column preparation by adding SDS just below its CMC to 1.5 g of alumina, then fine-tuning the pH to 2 with HCl before mixing for 10 min. The sample was washed with water then mixed with the organic ligand for 15 min., filtered and added to the column. The anion head group of the surfactant is bonded by charge attraction to the solid alumina, and the hydrophobic tail now becomes the external portion of the alumina particles giving a hydrophobic character to the phase and the organic ligand is adsorbed onto the treated alumina surface. In this fashion chromium (III) and total chromium were analyzed with an 8-hydroxyquinoline {CAS No. 134–31–6} as the immobilized organic reagent on a SDS-coated alumina column followed by flow injection atomic absorption spectrometry after elution with a 20 % ethanol solution in HCl. The linear calibration range was 1.0–100 µg/L and detection limit was 0.16 ng/L [175]. Silver was determined using a similar method with diethyldithiocarbamate {sodium (diethylcarbamothioyl)sulfanide, CAS No. 148–18–5} as the complexation agent immobilized on the SDS coated alumina column and analyzed by FAAS. An enrichment factor of 125 was achieved, and linear calibration was documented as 5–100 µg/L with a detection limit of 0.7 µg/L [176].

The analysis of pollutants, herbicides and pesticides has used chemiluminescence and fluorescence as the principle approaches for FIA investigations. The analysis of the phenylurea herbicide residues utilized FIA and fluorescence detection of the UV irradiated species of the herbicides in spiked tap water samples. Both SDS and CTAC were used at their CMC levels to solubilize the reaction products and fluorescence intensity enhancement, which exhibited linear calibration over two orders of magnitude with a detection limit of 0.33–0.92 mg/L [177]. The broadleaf herbicide metsulfuron-methyl, {2-{[(4-methoxy-6-methyl-1,3,5-triazin-2-yl)amino]-oxomethyl]sulfamoyl}benzoic acid methyl ester, CAS No. 74223-64–6}, a sulfonylurea, was analyzed in environmental waters using a solid phase spectroscopy system and photochemical induced fluorescence. The irradiated product occurred was retained in a flow cell filled with ODS, octadecylsilane, silica gel coated with SDS had a detection limit of 0.14 µg/L. The solid phase elution was with a 30 % methanol–water solvent with 8 mM SDS between runs. The size of the injection volume of 1000 µL regulated the analytical range of 0.5–175 µg/L [178].

The very clever use of CPE has also been coupled with FIA [179]. Silva et al. sequestered and analyzed lead and cadmium by FAAS after CPE preconcentration in a FIA system. The metal ions were chelated with TAN, {1-[2-(2-thiazolyl)diazenyl]-2-naphthalenol, CAS No. 1147–56–4}, which migrated to the micelles of Triton X-114 and were subsequently concentrated onto a cotton or glass wool mini–column and flushed off for determination by FAAS by the addition of HCl. The enhancement factors were between 15.1 and 20.3, and detection limits were 4.5 µg/L for Pb and 0.75 µg/L for Cd [180]. A similar system had been used for the determination of benzo[a]pyrene using the peroxoxalate chemiluminescence detection system. A cotton filled mini-column collected Triton X-114 CPE phase that had its cloud point temperature depressed by using sodium sulfate in the mobile phase. Elution was achieved using pure acetonitrile. The method was based on the chemiluminescence product that resulted from the reaction of hydrogen peroxide oxidation and bis(2,4,6-trichlorophenyl)oxalate {CAS No. 1165–91–9}. The enrichment factors recounted as 40.2 for benzo[a]pyrene, 38.5 for benzo[k]fluoranthene and 41.3 for benzo[ghi]perylene [181]. The detection limits were brought down to the picomolar and femtomolar concentration levels in an ultrasensitive determination of silver, gold, and iron oxide nanoparticles in environmental samples that combines preconcentration using CPE and the chemical luminescent oxidation of luminol {5-amino-2,3-dihydrophthalazine-1,4-dione, CAS No. 521-31-3} [182]. Luminol was used recently in the determination of L-thyroxine in pharmaceutical preparations along with CTAB and $KMnO_4$ that acted as enhancement factors for the chemiluminescence detection [183]. The FIA and chemiluminescence detection of gemifloxacin in pharmaceutical preparations and biological fluids used a CTAB surfactant system and the reaction of the analyte with diperiodatoargentate (III) [184].

6.5 Conclusions

This chapter has attempted to give an introduction to the many practical uses of surfactants and elucidate some background in our understanding of how they can be utilized to achieve green analytical chemistry. Seeing how a micellar system works can also give insight into new methods based on the solubilization, or catalytic properties of surfactant systems. The opportunity of both enhancement, and the replacement of organic solvents make micelles a solvent system to go to for cost and environmental purposes. A comprehensive look at micelles in analytical chemistry would require a separate encyclopedia and would be out of date right after it was written for this still very active field of research. All aspects of analytical chemistry are being explored, however, some fields such as, fluorescence and UV-Visible spectroscopy have been active since the 1970s. Other analytical methods offer new and exciting research uses for those same chemicals with which we wash our hands, our clothes and that we use every day inside and outside the chemistry laboratory (Table 6.2).

Table 6.2: Surfactant properties [2, 4, 48, 185, 186].

Surfactant (CAS No.)	GMM g/mole	Aggregation number	CMC mMolar	Formula
Aerosol-22 (3401-73-8)	653		0.70 [8]	$C_{26}H_{43}NO_{16}Na_4S$
Aerosol-OT, AOT (577-11-7)	444.56		3.2 [9]	$C_{20}H_{37}O_7NaS$
Brij 35 (9002-92-0)	1225	40	0.09	$C_{12}H_{25}O(CH_2CH_2O)_{23}H$
Brij 96 (9009-91-0)	709		0.94 [1]	$HO(CH_2CH_2O)_{10}C_{18}H_{35}$
CPB, Cetylpyridinium bromide (140-72-7)	402.47		0.90 [11]	$C_{21}H_{38}NBr{\cdot}H_2O$
CPC, Cetylpyridinium chloride (6004-24-6)	358.01	18-55	0.12 [6]	$C_{21}H_{38}NCl{\cdot}H_2O$
CTAB, Cetyl trimethylammonium bromide (9036-06-0)	364.5	170	0.92 [3]	$(C_{16}H_{33})N(CH_3)_3Br$
CTAC, Cetyl trimethylammonium chloride (79728-63-5)	320.0	84	1.3 [3]	$[(C_{16}H_{33})N(CH_3)_3Cl$
SDS, Sodium dodecyl sulfate (151-21-3)	288.38		8.2 [11]	$C_{12}H_{25}SO_4Na$
Triton X-100 (9036-19-5)	650	140	0.23 [12]	$C_8H_{17}C_6H_4(OC_2H_4)_{10}OH$
Triton X-114 (9081-83-8)	537			$C_8H_{17}C_6H_4O(CH_2CH_2O)_{7.5}H$
Tween-20 (93906-96-8)	1228		0.059 [5]	$C_{58}H_{114}O_{26}$
Tween-80 (9005-65-6)	1310	58	0.012 [5]	$C_{18}H_{37}$-$C_6H_9O_5$-$(OC_2H_4)_{20}OH$

References

[1] Birdi KS, Singh HN, Dalsager SU. Interaction of ionic micelles with the hydrophobic fluorescent probe 1–anilino–8–naphthalenesulfonate. J Phys Chem. 1979;83(21):2733–2737.
[2] Rosen MJ. Surfactants and interfacial phenomena. New York, NY: John Wiley & Sons, 1978.
[3] Atwood D, Florence AT. Surfactants systems their chemistry, pharmacy and biology. New York, NY: Chapman and Hall, 1983.
[4] Tadros TF. Surfactants. Orlando, FL: Academic Press, 1984.
[5] Urata K, Takaishi N. A perspective on the contribution of surfactants and lipids toward "Green Chemistry": present states and future potential. J Surfactants Deterg. 2001;4(2): 191–200.
[6] Aniansson GE. Dynamics and structure of micelles and other amphiphile structures. J Phys Chem. 1978; 82(26):2805–2808.
[7] Thomas JK. Radiation–induced reactions in organized assemblies. Chem Rev. 1980;80(4):283–299.
[8] Emert J, Behrens C, Goldenberg M. Intramolecular excimer–forming probes of aqueous micelles. J Am Chem Soc. 1979;101(3):771–772.
[9] Merrettig-Bruns U, Jelen E. Anaerobic biodegradation of detergent surfactants. Materials. 2009;2(1):181–206.
[10] Rosen MJ, Tracy DJ. Gemini surfactants. J Surfactants Deterg. 1998;1(4):547–554.
[11] Menger FM, Littau CA. Gemini surfactants: a new class of self–assembling molecules. J Am Chem Soc. 1993;115(22):10083–10090 and Menger FM, and Keiper JS. Gemini surfactants. Angewandte Chemie International Edition 2000, 39(11), 1906–20.
[12] Zana R. Dimeric (gemini) surfactants: effect of the spacer group on the association behavior in aqueous solution. J Colloid Interface Sci. 2002;248(2):203–220.
[13] Singh V, Tyagi R. Unique micellization and cmc aspects of gemini surfactant: an overview. J Dispersion Sci Technol. 2014;35(12):1774–1792.
[14] Kalyanasundaram K, Thomas JK. The conformational state of surfactants in the solid state and in micellar form A laser–excited Raman scattering study. J Phys Chem. 1976;80(13): 1462–1473.
[15] Goodling K, Johnson K, Lefkowitz L, Williams BW. The modern student laboratory: luminescent characterization of sodium dodecyl sulfate micellar solution properties. J Chem Educ. 1994;71(1):A8.
[16] Nakahara Y, Kida T, Nakatsuji Y, Akashi M. New fluorescence method for the determination of the critical micelle concentration by photosensitive monoazacryptand derivatives. Langmuir. 2005;21(15):6688–6695.
[17] Wandruszka RV. Luminescence of micellar solutions. Crit Rev Anal Chem. 1992;23(3):187–215.
[18] Garcia MD, Sanz–Medel A. Dye–surfactant interactions: a review. Talanta. 1986;33(3): 255–264.
[19] Schramm LL, Stasiuk EN, Marangoni DG. 2 Surfactants and their applications. Annu Rep Sect "C" (Phys Chem). 2003;99:3–48.
[20] Memon N, Balouch A, Hinze WL. Fluorescence in organized assemblies. Encycl Anal Chem. 2008;9:16-110.
[21] Gu T, Sjöblom J. Surfactant structure and its relation to the Krafft point, cloud point and micellization: some empirical relationships. Colloids Surf. 1992;64(1):39–46.
[22] Hinze WL, Pramauro E. A critical review of surfactant–mediated phase separations (cloud–point extractions): theory and applications. Crit Rev Anal Chem. 1993;24(2):133–177.
[23] Quina FH, Hinze WL. Surfactant–mediated cloud point extractions: an environmentally benign alternative separation approach. Ind Eng Chem Res. 1999;38(11):4150–4168.

[24] Kalyanasundaram K, Thomas JK. Environmental effects on vibronic band intensities in pyrene monomer fluorescence and their application in studies of micellar systems. J Am Chem Soc. 1977;99(7):2039–2044.
[25] Thomas JK. Effect of structure and charge on radiation–induced reactions in micellar systems. Acc Chem Res. 1977;10(4):133–138.
[26] Itoh H, Ishido S, Nomura M, Hayakawa T, Mitaku S. Estimation of the hydrophobicity in microenvironments by pyrene fluorescence measurements: N–β–octylglucoside micelles. J Phys Chem. 1996;100(21):9047–9053.
[27] Grätzel M, Thomas JK. Dynamics of pyrene fluorescence quenching in aqueous ionic micellar systems factors affecting the permeability of micelles. J Am Chem Soc. 1973;95(21):6885–6889.
[28] Piñeiro L, Novo M, Al–Soufi W. Fluorescence emission of pyrene in surfactant solutions. Adv Colloid Interface Sci. 2015;215:1–2.
[29] Fendler JH, Fendler EJ, Infante GA, Shih PS, Patterson L. Absorption and proton magnetic resonance spectroscopic investigation of the environment of acetophenone and benzophenone in aqueous micellar solutions. J Am Chem Soc. 1975;97(1):89–95.
[30] Drummond CJ, Grieser F, Healy TW. A single spectroscopic probe for the determination of both the interfacial solvent properties and electrostatic surface potential of model lipid membranes. Faraday Discuss Chem Soc. 1986;81:95–106.
[31] Karukstis KK, Suljak SW, Waller PJ, Whiles JA, Thompson EH. Fluorescence analysis of single and mixed micelle systems of SDS and DTAB. J Phys Chem. 1996;100(26):11125–11132.
[32] Mandal D, Pal SK, Bhattacharyya K. Excited–state proton transfer of 1–naphthol in micelles. J Phys Chem A. 1998;102(48):9710–9714.
[33] Almgren M, Grieser F, Thomas JK. Dynamic and static aspects of solubilization of neutral arenes in ionic micellar solutions. J Am Chem Soc. 1979;101(2):279–291.
[34] Ashby KD, Das K, Petrich JW. The effect of micelles on the steady–state and time–resolved fluorescence of indole, 1–methylindole, and 3–methylindole in aqueous media. Anal Chem. 1997;69(10):1925–1930.
[35] Martens FM, Verhoeven JW. Charge–transfer complexation in micellar solutions. Water penetrability of micelles. J Phys Chem. 1981;85(13):1773–1777.
[36] Infelta PP, Gratzel M, Thomas JK. Luminescence decay of hydrophobic molecules solubilized in aqueous micellar systems kinetic model. J Phys Chem. 1974;78(2):190–195.
[37] Marzzacco CJ. The effect of solvent and micelles on the rate of excited–state deprotonation. J Chem Educ. 1996;73(3):254.
[38] Grieser F, Drummond CJ. The physicochemical properties of self–assembled surfactant aggregates as determined by some molecular spectroscopic probe techniques. J Phys Chem. 1988;20:5580–5593.
[39] Hartly GS. The effect of long–chain salts on indicators: the valence–type of indicators and the protein error. Trans Faraday Soc. 1934;30:444–450.
[40] Duynstw EFJ, Grunwald E. Organic reactions occurring in or on micelles I. Reaction rate studies of the alkaline fading of triphenylmethane dyes and sulfonphthalein indicators in the presence of detergent salts. J Am Chem Soc. 1959;81(17):4540–4542.
[41] Reeves RL. Nature of mixed micelles from anionic dyes and cationic surfactants kinetic study. J Am Chem Soc. 1975;97(21):6019–6024.
[42] Reeves RL. Effect of reacting and competing counterions on the hydrolysis kinetics of an anionic dye ester in mixed micelles with CTAB [hexadecyltrimethylammonium bromide]. J Am Chem Soc. 1975;97(21):6025–6029.
[43] Triboni ER, Politi MJ, Cuccovia IM, Chaimovich H, Berci Filho P. Rate limiting step and micellar catalysis of the non-classical nitro group nucleophilic substitution by thiols in 4 nitro N n butyl 1, 8 naphthalimide. J Phys Org Chem. 2003;16(6):311–317.

[44] Katusin–Razem B, Wong M, Thomas JK. The effect of micellar phase on the state and dynamics of some excited state charge transfer complexes. J Am Chem Soc. 1978;100(6):1679–1686.
[45] Alkaitis SA, Beck G, Grätzel M. Laser photoionization of phenothiazine in alcoholic and aqueous micellar solution electron transfer from triplet states to metal ion acceptors. J Am Chem Soc. 1975;97(20):5723–5729.
[46] La Sorella G, Strukul G, Scarso A. Recent advances in catalysis in micellar media. Green Chem. 2015;17(2):644–683.
[47] Hinze WL.. Mittal KL, editors. Use of surfactant and micellar systems in analytical chemistry. Solution Chem. Surfactants. Proc. Sect. 52nd Colloid Surf. Sci. Symp., 1979;79–127 1.
[48] Cline–Love LJ, Habarta JG, Dorsey JG. The micelle–analytical chemistry interface. J Anal Chem. 1984;56(11) 1132A–1148A.
[49] Pelizzetti E, Pramauro E. Analytical applications of organized molecular assemblies. Anal Chim Acta. 1985;169:1–29.
[50] McLntire GL, Dorsey JG. Micelles in analytical chemistry. Crit Rev Anal Chem. 1990;21(4): 257–278.
[51] Fendler JH, Fendler EJ. Catalysis in micellar and micromolecular systems. New York: Academic Press, 1975.
[52] Lewis M, Wee V. Aquatic safety assessment for cationic surfactants. Environ Toxicol Chem. 1983;2:105–108.
[53] Schramm LL. Emulsions, foams and suspensions: fundamentals and applications. Hoboken, NJ: John Wiley & Sons, 2006 May 12. May 12.
[54] Schramm LL. Surfactants: fundamentals and applications in the petroleum industry. Cambridge, UK: Cambridge University Press, 2000 Mar 23. Mar 23.
[55] Scott MJ, Jones MN. The biodegradation of surfactants in the environment. Biochim Biophys Acta. 2000;1508(1–2):235–225.
[56] Ying GG. Fate, behavior and effects of surfactants and their degradation products in the environment. Environ Int. 2006;32(3):417–431.
[57] Olkowska E, Polkowska Z, Namieśnik J. Analytics of surfactants in the environment: problems and challenges. Chem Rev. 2011;111(9):5667–5700.
[58] Oei HH, Mai I, Toro DC. Quantitative analysis of anionic or cationic surfactants using a surfactant electrode. J Soc Cosmet Chem. 1991;42(5):309–316.
[59] Shah A, Shahzad S, Munir A, Nadagouda MN, Khan GS, Shams DF, et al. Micelles as soil and water decontamination agents. Chem Rev. 2016;116(10):6042–6074.
[60] Kim SO, Moon S–H, Kim KW. Removal of heavy metals from soils using enhanced electrokinetic soil processing. Water Air Soil Pollut. 2001;125 259–272.
[61] Jiang Y, Zhan H, Yuan J, Ma M, Chen H. Washing efficiency of heavy metals in soils with EDTA enhanced by surfactants. J Agric Res. 2006;25 119–123.
[62] Westall JC, Chen H, Zhang W, Brownawell BJ. Sorption of linear alkylbenzenesulfonates on sediment materials. Environ Sci Technol. 1999;33(18):3110–3118.
[63] Krueger CJ, Radakovich KM, Sawyer TE, Barber LB, Smith RL, Field JA. Biodegradation of the surfactant linear alkylbenzenesulfonate in sewage–contaminated groundwater: a comparison of column experiments and field tracer tests. Environ Sci Technol. 1998;32(24):3954–3961.
[64] McAvoy DC, Eckhoff WS, Rapaport RA. Fate of linear alkylbenzene sulfonate in the environment. Environ Toxicol Chem. 1993;12:977–987.
[65] Brunner PH, Capri S, Marcomini A, Giger W. Occurrence and behaviour of linear alkylbenzenesulphonates, nonylphenol, nonylphenol mono–and nonylphenol diethoxylates in sewage and sewage sludge treatment. Water Res. 1988;22(12):1465–1472.
[66] Ruiz BF, Prats D, Rico C. Elimination of LAS (linear alkylbenzene sulfonate) during sewage treatment, drying and composting of sludge and soil amending processes. In: Quaghebeur D,

Temmerman I, Angeletti G, editor(s). Organic contaminants in waste water, sludge and sediment. London: Elsevier Applied Science, 1989.
[67] Mieure JP, Waters J, Holt MS, Matthijs E. Terrestrial safety assessment of linear alkyl benzene sulphonate. Chemosphere. 1990;21:251–262.
[68] Hernández-Soriano MDC, Degryse F, Smolders E. Mechanisms of enhanced mobilisation of trace metals by anionic surfactants in soil. Environ Pollut. 2011;159(3):809–816.
[69] Hernández-Soriano MDC, Peña A, Mingorance MD. Release of metals from metal-amended soil treated with a sulfosuccinamate surfactant: effects of surfactant concentration, soil/solution ratio, and pH. J Environ Qual. 2010;39(4):1298–1305.
[70] Edwards DA, Adeel Z, Luthy RG. Distribution of nonionic surfactant and phenanthrene in a sediment/aqueous system. Environ Sci Technol. 1994;28:1550–1560.
[71] Brunner PH, Capri S, Marcomini A, Giger W. Occurrence and behavior of linear alkylbenzene sulphonates, nonylphenol, nonylphenol mono- and nonylphenol diethoxylates in sewage and sewage sludge treatment. Water Res. 1988;22:101–113.
[72] EPA-factsheet PFOA & PFOS Drinking Water Health Advisories US Environmental Protection Agency May. July 2016. EPA 800-F-16-003 Available at: https://www.epa.gov/ground-water-and-drinking-water/drinking-water-health-advisories-pfoa-and-pfos. Accessed: July2016.
[73] Soares A, Guieysse B, Jefferson B, Cartmell E, Lester JN. Nonylphenol in the environment: a critical review on occurrence, fate, toxicity and treatment in wastewaters. Environ Int. 2008;34(7):1033–1049.
[74] Renner R. European bans on surfactant trigger transatlantic debate. Environ Sci Technol. 1997;31(7):316A–320A.
[75] David A, Fenet H, Gomez E. Alkylphenols in marine environments: distribution monitoring strategies and detection considerations. Mar Pollut Bull. 2009;58(7):953–960.
[76] Aronstein BN, Calvillo YM, Alexander M. Effect of surfactants at low concentrations on the desorption and biodegradation of sorbed aromatic compounds in soil. Environ Sci Technol. 1991;25:1728–1731.
[77] Kosswig K. Surfactants. In: W. Gerhartz, editor(s). Ullmann's encyclopedia of industrial chemistry Vol. A25. Weinheim, Germany: VCH Verlagsgesellschaft mbH, 1994:747–816.
[78] Metcalfe TL, Dillon PJ, Metcalfe CD. Transport of toxic substances from golf courses into watersheds in the precambrian shield region of Ontario, Canada. Environ Toxicol Chem. 2008;27:811–818.
[79] Watanabe H. Spectrophotometric determination of cobalt with 1-(2-pyridylazo)-2-naphthol and surfactants. Talanta. 1974;21(4):295–302.
[80] Watanabe H, Sakai Y. Spectrophotometric determination of zinc with 1-(2-pyridylazo)- 2-naphthol and surfactant. Bunseki Kagaku. 1974;23(4):396–402 Chem. Abstr. 1974, 81, 32939b.
[81] Watanabe H, Miura J. Selection of masking agents in the spectrophotometric determination of nickel with 1-(2-thiazolylazo)-2-naphthol and nonionic surfactant. Bunseki Kagaku. 1976;25(10):667–670 Chem. Abstr. 1977, 87, 126585j.
[82] Watanabe H, Ishii H. Spectrophotometric determination of nickel with 2-(2-thiazolylazo)-5-dimethylaminophenol and Triton X-100. Bunseki Kagaku. 1977;26(2):86–91 Chem. Abstr. 1977, 87, 126574e.
[83] Hinze WL, Williams RW, Fu ZS, Suzuki Y, Quina FH. Novel chiral separation techniques based on surfactants. Colloids Surf. 1990;48:79–94.
[84] Ostlund SG, Pharr DY. Determination of cobalt(II) by flow injection analysis using a surfactant solvent system. J Undergraduate Chem Res. 2003;2(1):33–38.
[85] Marczenko Z, Jarosz M. Formation of ternary complexes of aluminium with some triphenylmethane reagents and cationic surfactants. Analyst. 1982;107(1281):1431–1438.

[86] Sharma R, Kamal A, Mahajan RK. A quantitative appraisal of the binding interactions between an anionic dye, Alizarin Red S, and alkyloxypyridinium surfactants: a detailed micellization, spectroscopic and electrochemical study. Soft Matter. 2016;12(6):1736–1749.
[87] Šimek J, Son NT, Ružička E. Spectrophotometry study of the reaction between dyes of the Alizarin green series and vanadates in the presence of cetylpyridinium cation. Chem Pap. 1985;39(1):91–101.
[88] Šimek J, Son NT, Ružička E. Spectrophotometric study of the reaction of Alizarin green series dyes with uranyl ions in the presence of cationoid surfactants. Collect Czechoslov Chem Commun. 1985;50(3):611–620.
[89] Tsao FP, Underwood AL. Spectrophotometric determination of nitrite with p–nitroaniline and 2–methyl–8–quinolinol in hexadecyl–trimethylammonium bromide solution. Anal Chim Acta. 1982;136:129–134.
[90] Nair J, Gupta VK. The spectrophotometric determination of nitrite in water with 8–quinolinol. Anal Chim Acta. 1979;111:311–314.
[91] Sarpal RS, Belletete M, Durocher G. Fluorescence probing and proton–transfer equilibrium reactions in water, SDS, and CTAB using 3, 3–dimethyl–2–phenyl–3H–indole. J Phys Chem. 1993;97(19):5007–5013.
[92] Spurlin S, Hinze W, Armstrong DW. Use of an aqueous micellar medium to improve the spectrophotometry determination of cyanide ion with 5, 5′–Dithiobis (2–Nitrobenzoic Acid). Anal Lett. 1977;10(12):997–1008.
[93] Berezin IV, Yatsimirskaya NT, Yatsimirskii AK, Martinek K. Use of micellar catalysis in nucleophilic substitution and addition in organic analysis. Dokl Akad Nauk SSSR. 1986;287(3):643–647.
[94] Connors KA, Wong MP. Micellar catalysis of an analytical reaction: spectrophotometric determination of amino acids and peptides after cetrimonium bromide catalyzed reaction with 1 fluoro 2, 4 dinitrobenzene. J Pharm Sci. 1979;68(11):1470–1471.
[95] Liu Y, Wu H, Ma W. Spectrophotometric determination of benzalkonium bromide in pharmaceutical samples with Alizarin green. J Surfactants Deterg. 2013;16(2):265–269.
[96] Watanabe H, Ohmori H. Dual–wavelength spectrophotometric determination of cadmium with cadion. Talanta. 1979;26(10):959–961.
[97] Fu–Sheng W, Fang Y. Spectrophotometric determination of silver with cadion 2B and triton X–100. Talanta. 1983;30(3):190–192.
[98] Hung SC, Qu CL, Wu SS. Spectrophotometric determination of uranium (VI) with 2–(3, 5–dibromo–2–pyridylazo)–5–diethylaminophenol in the presence of anionic surfactant. Talanta. 1982;29(7):629–631.
[99] Zhang CP, Qi DY, Zhou TZ. Sensitive spectrophotometric determination of traces of zirconium with 2–(6–bromo–2–benzothiazolylazo)–5–diethylaminophenol in the presence of sodium lauryl sulphate. Talanta. 1982;29(12):1119–1121.
[100] Hung SC, Qu CL, Wu SS. Spectrophotometric determination of silver with 2–(3, 5–dibromo–2–pyridylazo)–5–diethyl–aminophenol in the presence of anionic surfactant. Talanta. 1982;29(2):85–88.
[101] Zhe T, Wu SS. Spectrophotometric determination of zinc with 2–(3, 5–dibromo–2–pyridylazo)–5–diethylaminophenol in the presence of anionic surfactant. Talanta. 1984;31(8):624–626.
[102] Malik AK, Kaul KN, Lark BS, Faubel W, Rao AL. Spectrophotometric determination of cobalt, nickel palladium, copper, ruthenium and molybdenum using sodium isoamylxanthate in presence of surfactants. Turkish J Chem. 2001;25(1):99–105.
[103] Ghaedi M. Selective and sensitized spectrophotometric determination of trace amounts of Ni (II) ion using α–benzyl dioxime in surfactant media. Spectrochim Acta A Mol Biomol Spectrosc. 2007;66(2):295–301.

[104] Leon–Gonzalez ME, Santos–Delgado MJ, Polo–Diez LM. An improved method for the spectrophotometric determination of fluoride by addition of sodium dodecyl sulphate to the fluoride/lanthanum (III)/Alizarin fluorine blue system. Anal Chim Acta. 1985;178:331–335.
[105] Oshiro I, Takenaka T, Maeda J. New method for hemoglobin determination by using sodium lauryl sulfate (SLS). Clin Biochem. 1982;15(2):83–88.
[106] ThorburnáBurns D, Dagar D. Improvements to the spectrophotometric determination of germanium with phenylfluorone. Analyst. 1980;105(1246):75–79.
[107] Georges J. Molecular fluorescence in micelles and microemulsions: micellar effects and analytical applications. Prog Anal Spectrosc. 1990;13(1):27–45.
[108] Hinze WL, Singh HN, Baba Y, Harvey NG. Micellar enhanced analytical fluorimetry. Trac Trends Anal Chem. 1984;3(8):193–199, and Hinze WL, Srinivasan N, Smith TK, Igarashi S, and Hoshino H. Organized assemblies in analytical chemiluminescence spectroscopy: an overview. Advances in multidimensional luminescence 1990, 1, 149–206.
[109] Ishibashi N, Kina K. Sensitivity enhancement of the fluorometric determination of aluminum by the use of surfactant. Anal Lett. 1972;5(9):637–641.
[110] Sanz–Medel A, Fernandez Perez MM, De la Guardia Cirugeda M, Dominguez JC. Metal chelate fluorescence enhancement by nonionic micelles: surfactant and auxiliary ligand nature influence on the niobium–lumogallion complex. Anal Chem. 1986;58(11):2161–2166.
[111] Sanz–Medel A, de la Campa RF, Alonso JI. Metal chelate fluorescence enhancement in micellar media: mechanisms of surfactant action. Analyst. 1987;112(4):493–497.
[112] Dominguez JC, de la Guardia Cirugeda M. Some observations on the fluorometric determination of metallic elements in micellar media. Microchemical Journal. 1989;39(1):50–58.
[113] Singh H, Hinze WL. Micellar enhanced spectrofluorometric methods: application to the determination of pyrene. Anal Lett. 1982;15(A3):221–243.
[114] Singh HN, Hinze WL. Micellar enhanced fluorimetric determination of 1–NN–dimethylamino-naphthalene–5–sulphonyl chloride and o–phthalaldehyde–2–mercaptoethanol derivatives of amino acids. Analyst. 1982;107(1278):1073–1080.
[115] Medina J, Hernandez F, Marin R, Lopez FJ. Study of the fluorescence of the lead–morin system in the presence of nonionic surfactants. Analyst. 1986;111(2):235–237.
[116] Taketatsu T. Complex formation of europium with thenoyltrifluoroacetone and tri–n–octylphosphine oxide in micellar solution of nonionic surfactant. Chem Lett. 1981; 10(8):1057–1058.
[117] Xu YY, Hemmilä IA. Co–fluorescence enhancement system based on pivaloyltrifluoroacetone and yttrium for the simultaneous detection of europium, terbium, samarium and dysprosium. Anal Chim Acta. 1992;256(1):9–16.
[118] Xu YY, Hemmilä IA, Lövgren TN. Co–fluorescence effect in time–resolved fluoroimmunoassays – A review. Analyst. 1992;117(7):1061–1069.
[119] Galdú MV, de la Guardia M, Braco L. Influence of anionic surfactants on the sensitisation of the fluorimetric determination of fenproporex. Analyst. 1987;112(7):1047–1050.
[120] De la Guardia Cirugeda M, Soriano FR. Micellar enhancement of benzodiazepine fluorescence. Analyst. 1989;114(1):77–82.
[121] Jee RD. Study of micellar solutions to enhance the europium–sensitized luminescence of tetracyclines. Analyst. 1995;120(12):2867–2872.
[122] Campaña AM, Barrero FA, Ceba MR. Sensitive spectrofluorimetric method for the determination of ethylenediaminetetraacetic acid and its salts in foods with zirconium ions and Alizarin Red S in a micellar medium. Anal Chim Acta. 1996;329(3):319–325.
[123] Diaz Garcia ME, Sanz–Medel A. Facile chemical deoxygenation of micellar solutions for room temperature phosphorescence. Anal Chem. 1986;58(7):1436–1440.

[124] Kalyanasundaram K, Grieser F, Thomas JK. Room temperature phosphorescence of aromatic hydrocarbons in aqueous micellar solutions. Chem Phys Lett. 1977;51(3):501–505.
[125] Love LC, Skrilec M, Habarta JG. Analysis of micelle–stabilized room temperature phosphorescence in solution. Anal Chem. 1980;52(4):754–759.
[126] Skrilec M, Love LC. Micelle–stabilized room–temperature phosphorescence characteristics of carbazole and related derivatives. J Phys Chem. 1981;85(14):2047–2050.
[127] Love LC, Grayeski ML, Noroski J, Weinberger R. Room–temperature phosphorescence, sensitized phosphorescence and fluorescence of licit and illicit drugs enhanced by organized media. Anal Chim Acta. 1985;170:3–12.
[128] Carretero AS, Blanco CC, Fernández RE, Gutiérrez AF. Micellar–stabilized room–temperature phosphorimetric determination of the fungicide thiabendazole in canned pineapple samples. Fresenius' J Anal Chem. 1998;360(5):605–608.
[129] Sanz–Medel A, Martinez Garcia PL, Diaz Garcia ME. Micelle–stabilized room–temperature liquid phosphorimetry of metal chelates and its application to niobium determination. Anal Chem. 1987;59(5):774–778.
[130] Liu YM, de la Campa MR, García ME, Sanz–Medel A. Micelle–stabilized liquid room–temperature phosphorimetry for metals: the micellar reaction of gallium with 7–iodo–hydroxyquinoline–5–sulfonic acid and its application to the metal determination. Microchimica Acta. 1991;103(1–2):53–64.
[131] Fendler JH, Gilbert RD, Yen TF. Advances in the applications of membrane-mimetic chemistry. New York, NY: Springer Science & Business Media, 2012 Dec 6. 79–94. Dec 6.
[132] Hickman A, Pharr DY. Enhancement effects of surfactants on the DC plasma analysis of P-block elements (in three parts). J Undergraduate Chem Res. 2012;11(2):54–58 and part II: 11(4), 102–106;and part III: 2013, 12(2), 28–32.
[133] Armstrong DW, McNeely M. Use of micelles in the TLC separation of polynuclear aromatic compounds and amino acids. Anal Lett. 1979;12(12):1285–1291.
[134] Armstrong DW, Terrill RQ. Thin layer chromatographic separation of pesticides, decachlorobiphenyl, and nucleosides with micellar solutions. Anal Chem. 1979;51(13):2160–2163.
[135] Armstrong DW, Henry SJ. Use of an aqueous micellar mobile phase for separation of phenols and polynuclear aromatic hydrocarbons via HPLC. J Liq Chromatogr. 1980;3(5): 657–662.
[136] Armstrong DW, Nome F. Partitioning behavior of solutes eluted with micellar mobile phases in liquid chromatography. Anal Chem. 1981;53(11):1662–1666.
[137] Armstrong DW. Micelles in separations: practical and theoretical review. Sep Purif Methods. 1985;14(2):213–304.
[138] Borgerding MF, Williams RL, Hinze WL, Quina FH. New perspectives in micellar liquid chromatography. J Liq Chromatogr. 1989;12(8):1367–1406.
[139] Esteve–Romero J, Albiol–Chiva J, Peris–Vicente J. A review on development of analytical methods to determine monitorable drugs in serum and urine by micellar liquid chromatography using direct injection. Anal Chim Acta. 2016;926:1–6.
[140] Terabe S, Otsuka K, Ichikawa K, Tsuchiya A, Ando T. Electrokinetic separations with micellar solutions and open–tubular capillaries. Anal Chem. 1984;56(1):111–113.
[141] Khaledi MG. Micelles as separation media in high–performance liquid chromatography and high–performance capillary electrophoresis: overview and perspective. J Chromatogr A. 1997;780(1):3–40.
[142] Grossman PD, Colburn JC. Capillary electrophoresis: theory and practice. San Diego, CA: Academic Press, 2012 Dec 2. 159–187. Dec 2.
[143] Sieradzka E, Witt K, Milnerowicz H. The application of capillary electrophoresis techniques in toxicological analysis. Biomed Chromatogr. 2014;28(11):1507–1513.

[144] Timerbaev AR. Element speciation analysis using capillary electrophoresis: twenty years of development and applications. Chem Rev. 2012;113(1):778–812.
[145] Altria KD. Analysis of pharmaceuticals by capillary electrophoresis. New York and Philadelphia, PA:Springer Science & Business Media, 2013 Apr 17. Apr 17.
[146] Silva M. Micellar electrokinetic chromatography: a review of methodological and instrumental innovations focusing on practical aspects. Electrophoresis. 2013;34(1):141–158.
[147] Rusling JF. Electrochemistry in micelles, microemulsions and related microheterogeneous fluids. Electroanalytical chemistry. 1993, 18, 1–80. In: Bard AJ, Decker M, editor(s). Electroanalytical chemistry a series of advances vol. 18. Boca Raton, FL: CRC Press, 1994.
[148] Proske GE. New technique for polarography of water–insoluble compounds. Anal Chem. 1952;24(11):1834–1836.
[149] McIntire GL, Chiappardi DM, Casselberry RL, Blount HN. Electrochemistry in ordered systems 2 electrochemical and spectroscopic examination of the interactions between nitrobenzene and anionic, cationic, and nonionic micelles. J Phys Chem. 1982;86(14):2632–2640.
[150] Vittal R, Gomathi H, Kim KJ. Beneficial role of surfactants in electrochemistry and in the modification of electrodes. Adv Colloid Interface Sci. 2006;119(1):55–68.
[151] Vittal R, Gomathi H, Rao GP. Influence of a cationic surfactant on the modification of electrodes with nickel hexacyanoferrate surface films. Electrochimica Acta. 2000;45(13):2083–2093.
[152] Silva MF, Cerutti ES, Martinez LD. Coupling cloud point extraction to instrumental detection systems for metal analysis. Microchimica Acta. 2006;155(3–4):349–364.
[153] Paleologos EK, Giokas DL, Karayannis MI. Micelle–mediated separation and cloud–point extraction. Trac Trends Anal Chem. 2005;24(5):426–436.
[154] Ferrera ZS, Sanz CP, Santana CM, Rodrıguez JJ. The use of micellar systems in the extraction and pre–concentration of organic pollutants in environmental samples. Trac Trends Anal Chem. 2004 Aug 31;23(7):469–479 Aug31.
[155] Stalikas CD. Micelle–mediated extraction as a tool for separation and preconcentration in metal analysis. Trac Trends Anal Chem. 2002;21(5):343–355.
[156] Mukherjee P, Padhan SK, Dash S, Patel S, Mishra BK. Clouding behaviour in surfactant systems. Adv Colloid Interface Sci. 2011;162(1):59–79.
[157] Tani H, Kamidate T, Watanabe H. Micelle–mediated extraction. J Chromatogr A. 1997;780(1):229–241.
[158] Bordier C. Phase separation of integral membrane proteins in Triton X–114 solution. J Biol Chem. 1981;256(4):1604–1607.
[159] Underwood AL. Dissociation of acids in aqueous micellar systems. Anal Chim Acta. 1982;140(1):89–97.
[160] Fernandez MS, Fromherz P. Lipoid pH indicators as probes of electrical potential and polarity in micelles. J Phys Chem. 1977;81(18):1755–1761.
[161] Siggia S, Hanna JG. Quantitative organic analysis via functional groups. New York, NY: John Wiley & Sons, 1979.
[162] Underwood AL. Acid–base titrations in aqueous micellar systems. Anal Chim Acta. 1977;93:267–273.
[163] Starościk R, Maskiewicz E, Malecki F. Acid – Base titrations of barbiturates in aqueous micellar media. Fresenius' Z Anal Chem. 1987;329(4):472–474.
[164] Gerakis AM, Koupparis MA, Efstathiou CE. Micellar acid – base potentiometric titrations of weak acidic and/or insoluble drugs. J Pharm Biomed Anal. 1993;11(1):33–41.
[165] Ruzicka J, Hansen EH. Retro-review of flow-injection analysis. Trac Trends Anal Chem. 2008;27(5):390–393.
[166] Růžička J. The second coming of flow-injection analysis. Anal Chim Acta. 1992;261(1):3–10.

[167] Pharr DY. A review of the use of surfactants in flow injection analysis. Anal Lett. 2011;44(13):2287–2311.

[168] Memon N, D Tzanavaras P. Utilization of organized surfactant assemblies as solvents in flow injection analysis with emphasis to automated derivatization of organic analytes. Curr Anal Chem. 2014;10(3):338–348.

[169] Resing JA, Measures CI. Fluorometric determination of Al in seawater by flow injection analysis with in–line preconcentration. Anal Chem. 1994;66(22):4105–4111.

[170] Pharr DY, Tomsyck JA. Flow injection analysis of benzoyl peroxide using N, N, N, N-tetramethyl-p-phenylenediamine (TMPDA) and surfactants. Anal Lett. 2009;42(5):821–832.

[171] Pharr DY, Sienerth KD, Tutor MJ. The analysis of nickel and cobalt by flow injection analysis using PAR, 4-(2-pyridylazo)-resorcinol, and Triton X-100. J Und Chem Res. 2004; 3(2):57–63.

[172] Safavi AF, Mirzaee M, Hormozi Nezhad MR, Saghir N. Kinetic spectrophotometric determination of copper by flow injection analysis in cationic micellar medium. Spectrosc Lett. 2005; 38(1):13–22.

[173] Long F, Zhang X, Zou Y, Xia X, Jiang R, Zhang W. Determination of trace zinc in seawater using reverse FIA spectrophotometric method. Pige Kexue Yu Gongcheng. 2006;16(6):44–46.

[174] Ruengsitagoon W, Chisvert A, Liawruangrath S. Flow injection spectrophotometric determination of lead using 1, 5-diphenylthiocarbazone in aqueous micellar. Talanta. 2010;81(1): 709–713.

[175] Ahmadi SH, Shabani AM, Dadfarnia S, Taei M. On-line preconcentration and speciation of chromium by an 8-hydroxyquinoline microcolumn immobilized on surfactant-coated alumina and flow injection atomic absorption spectrometry. Turkish J Chem. 2007;31(2):191–199.

[176] Dadfarnia S, Shabani AH, Gohari M. Trace enrichment and determination of silver by immobilized DDTC microcolumn and flow injection atomic absorption spectrometry. Talanta. 2004;64(3):682–687.

[177] Irace Guigand S, Leverend E, Seye MD, Aaron JJ. A new on line micellar enhanced photochemically induced fluorescence method for determination of phenylurea herbicide residues in water. Luminescence. 2005;20(3):138–142.

[178] Flores JL, de Córdova ML, Am D. Flow-through optosensing device implemented with photochemically-induced fluorescence for the rapid and simple screening of metsulfuron methyl in environmental waters. J Environ Monit. 2009;11(5):1080–1085.

[179] Carabias-Martınez R, Rodrıguez-Gonzalo E, Moreno-Cordero B, Pérez-Pavón JL, Garcıa-Pinto C, Laespada EF. Surfactant cloud point extraction and preconcentration of organic compounds prior to chromatography and capillary electrophoresis. J Chromatogr A. 2000;902(1):251–265.

[180] Silva EL, dos Santos Roldan P. Simultaneous flow injection preconcentration of lead and cadmium using cloud point extraction and determination by atomic absorption spectrometry. J Hazard Mater. 2009;161(1):142–147.

[181] Song GQ, Lu C, Hayakawa K, Lin JM. Comparison of traditional cloud-point extraction and online flow-injection cloud-point extraction with a chemiluminescence method using benzo[a] pyrene as a marker. Anal Bioanal Chem. 2006;384(4):1007–1012.

[182] Tsogas GZ, Giokas DL, Vlessidis AG. Ultratrace determination of silver, gold, and iron oxide nanoparticles by micelle mediated preconcentration/selective back-extraction coupled with flow injection chemiluminescence detection. Anal Chem. 2014;86(7):3484–3492.

[183] Cao J, Wang H, Liu Y. Determination of l-thyroxine in pharmaceutical preparations by flow injection analysis with chemiluminescence detection based on the enhancement of the luminol–$KMnO_4$ reaction in a micellar medium. Spectrochim Acta A Mol Biomol Spectrosc. 2015;140:162–165.

[184] Zhao F, Zhao WH, Xiong W. Chemiluminescence determination of Gemifloxacin based on diperiodatoargentate (III) sulphuric acid reaction in a micellar medium. Luminescence. 2013;28(2):108–113.
[185] Perkowski J, Mayer J, Ledakowicz S. Determination of critical micelle concentration of non-ionic surfactants using kinetic approach. Colloids Surfaces A. 1995;101:103–106.
[186] Hait SK, Moulik SP. Determination of critical micelle concentration (CMC) of nonionic surfactants by donor-acceptor interaction with Iodine and correlation of CMC with hydrophile-lipophile balance and other parameters of the surfactants. J Surf Det. 2001;4(3):303–309.

David Consiglio

7 Biofuels, fossil energy ratio, and the future of energy production

Abstract: Two hundred years ago, much of humanity's energy came from burning wood. As energy needs outstripped supplies, we began to burn fossil fuels. This transition allowed our civilization to modernize rapidly, but it came with heavy costs including climate change. Today, scientists and engineers are taking another look at *biofuels* as a source of energy to fuel our ever-increasing consumption.

Keywords: biofuels, fossil fuels, sustainability

7.1 Introduction

If you turn on the news or read about current events, you will likely find many stories about climate change. Most of them read the same way:

> The world faces a crisis. Climate change is altering our planet in numerous and highly undesirable ways. Farmland is turning to desert, sea levels are rising, corals are bleaching, and storms may increase in severity and number. The main culprit is carbon dioxide that spews from our power plants and cars when they burn fossil fuels. The problems our planet faces will only worsen . . . that is, unless we do something to stop it.

So why don't we? Why don't we just stop polluting and reverse climate change?

In this chapter, you will learn about the basic chemistry and synthesis of biofuels, an alternative to fossil fuels. In addition, we will introduce the *fossil energy ratio*, or FER, a concept that makes it easier to understand alternative fuels and can point the way to improvements that can help make alternative energy a reality.

7.1.1 What are biofuels?

Biofuels were humanity's first supply of energy outside of our bodies. Beginning several hundred thousand years ago [1], humanity began to control and use fire. This innovation allowed humans to transform their world and way of living in ways no other animal ever had. Humans used fire to repel predators, drive prey, and stay warm. Over time, humans used fire to clear fields for planting, create metal from ore, and build our modern civilization.

Throughout this entire time period, humans burned *biofuels*. A biofuel is simply any organically-derived substance that can be burned to produce energy. The most

https://doi.org/10.1515/9783110445923-007

commonly known biofuel is wood, but humans around the planet have used all sorts of biofuels, including animal dung, animal fat, and grasses.

But, biofuels had shortcomings. They were not always readily available. Furthermore, as civilization progressed toward industrialization, more and more fuel was needed to drive the machinery of modern life. Whole regions were deforested in the quest to provide biofuels for our growing industry.

7.1.2 The transition to fossil fuels

Around 200 years ago, humans started shifting from burning biofuels to burning *fossil fuels*. Fossil fuels are ancient biofuels that have been fossilized. The fossil fuels in use today are coal (primarily fossilized trees), oil (primarily fossilized microscopic sea creatures), and natural gas (primarily the gas formed by the partial decomposition of many organic materials). As industry grew, fossil fuels took on a greater role; today they account for over 80 % of energy production [2].

For decades, fossil fuels powered the industrialization of the world, but fossil fuels are not without problems. Most of the easily obtained fossil fuels are gone – they were the first to be mined and drilled for. More pressingly, the fossilized carbon in these fuels stored has been released into the atmosphere, significantly increasing the amount of carbon dioxide in our air. Direct measurements of CO_2 levels in the air have been measured directly since the 1950s, when levels were around 315 ppm. At the time of this writing, they are just over 400 ppm and steadily rising [3]. Higher CO_2 levels result in an increase in global temperatures [4].

The effects of this warming have, until recently, been relatively minor. But, as CO_2 levels have increased more dramatically, worrying changes have started happening. Land ice is melting at higher rates, leading to rising sea levels [5]. Storms and severe weather have increased in frequency and intensity [6]. CO_2 concentration in the oceans is rising, leading to ocean acidification and the destabilization of coral reef ecosystems [7].

These effects could get significantly worse: sea level rise alone could result in the forced relocation of nearly a billion people [8]. Scientists are looking for ways to reduce the rate of CO_2 emission and reduce, if not fully reverse, the effects of climate change brought about by the industrial age.

In my first year high school chemistry course, students participate in a project entitled "Saving Miami" [9]. The premise is simple: if humanity continues to burn fossil fuels at projected rates, sea level rise will submerge much of the greater Miami area – millions of people will lose their homes. Students choose from one of seven hypothetical "solutions" that sequester carbon dioxide. Stoichiometry plays a central role in the project as students have to perform multiple conversions to calculate the costs, land use, and time required to "save" Miami using each of these solutions. The project does not have a correct answer. Instead, students have to make value judgments to justify their choice. Additionally, students work extensively with scientific notation in order to handle the enormous numbers associated with carbon dioxide emission. Students write a scientific paper, present their work, and participate in interviews to demonstrate their mastery of the content. Along the way, they gain a new understanding of the vast scale of the problem that climate change poses to our planet.

7.1.3 Fossil carbon versus atmospheric carbon

After 200 years of decreasing importance, biofuels have emerged as a possible weapon in the fight against climate change. But how do biofuels work, and how can they reduce CO_2 emissions?

All combustion, whether of a fossil fuel or biofuel, proceeds via the following basic formula:

$$C_xH_yO_z + O_2 \Rightarrow CO_2 + H_2O$$

In this equation, $C_xH_yO_z$ represents any organic molecule. As you can see, CO_2 is produced regardless of the fuel being burned. So how can biofuels *reduce* CO_2 emissions?

All fuels derive their carbon from photosynthesis. In other words, the carbon in wood, coal, even animal dung, was once in the air. Plants absorbed carbon dioxide from the air and turned it into complex organic molecules. In some cases, animals ate these plants and further transformed the carbon-containing molecules.

Under most circumstances, this carbon would return to the air when the plant or animal died and decayed. So when we burn a biofuel, the carbon that is released was, until very recently, already in the air. If we burned nothing but biofuels, the amount of CO_2 in the air would remain relatively constant – plants would absorb CO_2 through photosynthesis, and we would return that CO_2 to the air through combustion.

But when we burn a fossil fuel, the CO_2 that is released was removed from the air millions of years ago. Ancient plants and animals were sometimes buried underground and fossilized. This removed the carbon from the atmosphere. When we burn a fossil fuel, this ancient CO_2 is returned to the air.

This is why biofuels have taken on renewed interest – they can be burned without significantly contributing to atmospheric CO_2 levels. But how are biofuels produced, and how can they replace existing fossil fuels?

7.2 Solid biofuels

Solid biofuels are used in a similar way to solid fossil fuels. Wood, grass, farm, or forestry wastes are typically burned as they are. This can be done in facilities similar to coal plants, and, in most cases, existing coal plants can be modified to allow the use of solid biofuels [10].

In some cases, solid biofuels must be processed before they are burned. The most common kind of processing is known as *densifying*, which refers to the process of compressing particulate biomass (such as sawdust) into logs or pellets for easier transportation and burning. Densified biomass is often bound using tree sap or other forestry waste – in this way, an additional waste product can be burned, increasing the energy output of a solid biofuel.

7.3 Liquid biofuels

Unlike solid biofuels, the chemistry of liquid biofuels is complex and varied. Liquid fuels are used almost exclusively for transportation, but unlike coal and natural gas power plants, gasoline and diesel engines are far less flexible in terms of the fuels they can burn [11]. In order to replace fossil fuels in cars and trucks, liquid biofuels must closely mimic gasoline or diesel fuel.

7.3.1 Gasoline substitutes

There are several liquid fuels that can stand in for petroleum-derived gasoline, but by far the most common is *ethanol*. Ethanol, the kind of alcohol that humans drink, is generally produced via fermentation of grain. This process is very similar to the production of alcohol for drinking and follows these basic steps:

1. Grain is harvested and processed to remove non-fermentable material like stalks and leaves.
2. The grain is ground into a powder.
3. Enzymes are added to break down complex sugars and starches to form simple sugars.
4. Yeast and water are added.
5. The yeast ferments the sugar in the grain to form alcohol via the following equation:

$$C_6H_{12}O_6 \rightarrow 2C_2H_5OH + 2CO_2$$
$$\text{Glucose} \rightarrow \text{Ethanol} + \text{Carbon Dioxide}$$

1. The resulting mixture, known as *industrial beer*, is distilled to remove most of the water.
2. The remaining water is removed using molecular sieves.
3. The ethanol is denatured, typically with gasoline, to render it undrinkable.
4. The CO_2 is captured and used for carbonating beverages.
5. The solid by-product is sold to farmers as animal feed.

In the United States, all gasoline engines must be capable of burning a gasoline mixture containing a small percentage of ethanol [12]. In recent years, FlexFuel vehicles have been produced that can burn up to 85 % ethanol. In some countries, 100 % ethanol vehicles have been manufactured.

7.3.2 Diesel substitutes

Diesel fuel substitute, known as *biodiesel*, is derived primarily from vegetable oil, but can be made from any oil or fat source, including used cooking oil. The most common process by which vegetable oil is transformed into biodiesel is known as *transesterification*. Vegetable oil consists of three long, fatty-acid chains connected to a glycerol backbone. In transesterification, these three fatty-acid chains are chemically removed and converted into methyl esters as shown in Figure 7.1.

The catalyst for this reaction is typically sodium or potassium hydroxide.

Biodiesel is very similar to petroleum-derived diesel fuel and the two fuels can be mixed in any ratio [13]. There are, however, two important differences regarding biodiesel:

1. Biodiesel is an excellent solvent that is capable of dissolving some fuel lines. Any vehicle that uses biodiesel must use fuel lines that are insoluble in biodiesel.
2. Biodiesel has a higher *cloud point* than petroleum-derived diesel [14]. Cloud point refers to the temperature at which the liquid becomes cloudy. At temperatures below the cloud point, biodiesel can become too viscous to flow easily into the engine. As a result, biodiesel is easier to use in warmer climates.

7.4 Gaseous biofuels

Many organic materials can be converted into methane using either chemical or biological processes [10]. This methane is often impure, containing significant quantities of carbon dioxide and other gases. This impure mixture can be burned

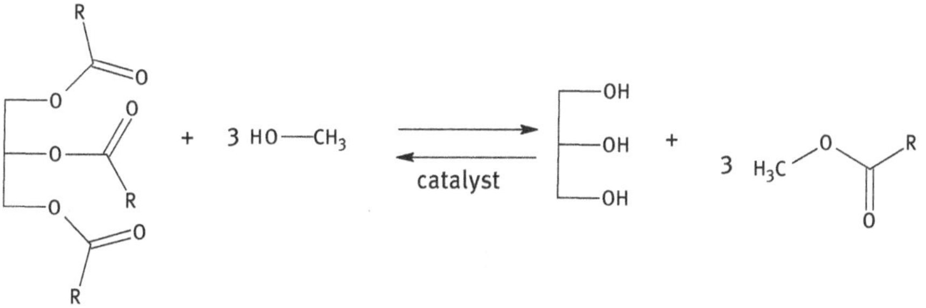

Figure 7.1: Transesterification.

as a lower-quality fuel, or purified and used in the existing natural gas infrastructure. In addition, other flammable gases can be produced from organic materials.

7.4.1 Biomethane and methane substitutes

Currently, there are two technologies that can be used to produce gaseous biofuels from solid or liquid biomass – digestion and gasification.

Digestion refers to the decomposition of solid or liquid biomass resulting in the production of methane. This process typically proceeds as follows:

$$C_{2x}H_yO_z \rightarrow xCO_2 + xCH_4$$

As you can see, this process does produce some CO_2 as a by-product, which reduces its impact on climate change.

Digestion is typically accomplished using anaerobic microorganisms that consume biomass and produce methane as a waste product. This process takes place at many landfills today, and in some cases the methane produced is used for power or heat generation.

Gasification involves drying and heating solid biomass to drive off water, and then hydrogen and oxygen, to make solid carbon called *char*. This char then reacts with water:

$$C + H_2O \rightarrow CO + H_2$$

This mixture of carbon monoxide and hydrogen gas is called *syngas* and can be burned in industrial power plants [15]. Syngas has a significant downside – carbon monoxide is highly toxic and must be used with extreme caution. It is not suitable for in-home use. Alternatively, the CO can be reacted with extra water to produce methane:

$$4CO + 2H_2O \rightleftharpoons CH_4 + 3CO_2$$

via a multistep process:

$$\begin{array}{c} CO + 3H_2\,CH_4 + H_2O \\ +\,3CO + 3H_2\,O3CO_2 + 3H_2 \\ \hline 4CO + 2H_2OCH_4 + 3CO_2 \end{array}$$

In my high school AP Chemistry course, I use these reactions as an example of an industrially important reversible reaction that takes place in multiple steps. The equilibrium constant of the overall reaction is the product of the equilibrium constants of the component reactions. In this case, the overall reaction is highly desirable but has a low equilibrium constant. Students use this information to design reaction conditions that maximize the yield of the desired product, in this case methane.

7.5 Fuel energy ratio

Every biofuel has a *fuel energy ratio*. This is simply the amount of energy a fuel produces divided by the amount of fossil energy it took to make that fuel:

$$FER = \frac{E_{out}}{E_{in}}$$

If a fuel has a high FER, then that fuel takes very little fossil fuel to make a large amount of energy. In contrast, fuels with a low FER take more fossil fuel energy to make, and produce less energy. Even fossil fuels have an FER – it takes fossil fuel energy to extract fossil fuel energy from the ground.

Clearly, a high value of FER is desirable in a fuel. But, how are energy output and energy input calculated?

7.5.1 Energy output

Calculating the energy output of a fuel is a simple matter of burning a unit of the fuel and measuring the energy given off by that fuel. This can be done via *calorimetry*. In this process, fuel is combusted with pure oxygen in a device known as a bomb calorimeter. The bomb portion of the calorimeter is submerged in water. After the fuel is burned, the temperature of the water will increase, as will the temperature of the calorimeter itself. Calculating the energy released by the fuel (ΔH) is then the simple matter of multiplying the change in temperature (ΔT) by the heat capacity of the calorimeter/water combination (C_v):

$$\Delta H = C_v \Delta T$$

The heat can then be divided by the amount of fuel that was burned, which gives the *heat of combustion* (ΔH_c). Scientists will typically calculate this value in kJ/mol, but industrially the units tend to be more practical and specific to the country in question. For example, the U.S. Government uses Btu/gal to measure energy output from liquid fuels.

In the end, the units do not matter so long as we are consistent: FER is a unitless quantity and FER values for different fuels can be directly compared.

Some scientists include the energy output values for co-products and by-products. For example, when corn ethanol is produced, there are leftover stalks, leaves, and other plant products that cannot be used to produce ethanol [16]. However, these can be burned to produce energy; therefore, some scientists include the energy output of these products in their calculations. It is important to determine if this energy has been used in the FER calculation before directly comparing FER values.

7.5.2 Energy input

Unlike energy output, which is straightforward to calculate, energy input is more difficult to determine precisely. This is due to the fact that energy inputs come from four major sources:

1. The energy required to grow the fuel crop – this includes fertilizer, insecticides, and energy to run sprinklers and water pumps
2. The energy required to plant and harvest the crop – this includes fuel for plows, tractors, and combines
3. The energy required to transport the crop for processing – this includes fuel for trucks and trains
4. The energy required to process the crop into fuel – this includes electricity for the processing plant, necessary chemical inputs for the crop, and disposal of waste materials

As you can imagine, these numbers take a bit more work to calculate. The details of energy inputs can be a source of considerable variability in FER values, and there is difficulty in determining some of them accurately. There are a few details worth noting:

1. Crop yields have been increasing steadily for decades as better farming practices, pesticides, fertilizer, and genetically engineered crops have increased the productivity of many fuel crops.
2. Processing crops into fuels near the fields where the crops are grown can reduce the energy costs of crop transportation.
3. Improvements in processing have reduced the production of waste, and uses for waste have arisen, both of which have reduced energy inputs for waste disposal.

7.5.3 FER values

Below is a list of values of FER for various fuels [17–25]:

Electrical Generation	Energy Ratio	Transportation Fuel	Energy Ratio
Coal – U.S. average	9	Cellulosic ethanol	5.4
Unscrubbed Western coal	6	Biodiesel	3.2–5.5
Biomass direct combustion	6–13	Wheat straw ethanol	1.6
Biomass gasification	2–5	Corn ethanol	0.85 – 2.00
Scrubbed Western coal	2.5	Petroleum-derived diesel	0.84
Natural gas fired	2.3	Gasoline	0.74–0.91

As you can see, there is a spectrum of FER values. Also, notice that FER is actually less than 1 for both diesel fuel and gasoline. This value results from the fact that while diesel fuel and gasoline emit a significant amount of energy when burned, extracting oil from the ground and refining that oil into diesel fuel or gasoline requires even more energy per unit. For comparison, consider that unrefined petroleum has a fuel energy ratio around 10. The net result is that both diesel fuel and gasoline produce less energy than was required to make them. In addition, many of the fuels have a range of FER values. This is due to the variability of production techniques, crop yields, and improvements in technology.

Coal has the highest energy ratio of any fuel currently used. This is due to the fact that coal is very abundant, relatively easy to mine, and simple to process. This makes replacing coal as a fuel more economically challenging than replacing gasoline and diesel fuel.

High school students often struggle with nuanced explanations. The high energy ratio of coal provides an opportunity to walk students through such an explanation. Coal truly is the cheapest fuel available – provided one ignores the cost of the pollution it produces. But if the cost of the pollution is considered, the relative value plummets. These two ways of looking at coal provide students with an insight into how facts and information can be skewed to support or refute various positions. By judging coal from the perspective of energy ratios (it is a wonderful fuel) and pollution (it is a terrible fuel), students can begin to step away from simplistic "good" and "bad" explanations and toward a deeper understanding of the challenges of modern life.

7.6 Problems with biofuels

Biofuels are not without their limitations. First among these is the staggering amount of fossil fuel used each year – far more than can be replaced using biofuels.

As an example, the United States harvested 86.6 million acres of corn in 2016 [26] at an average yield of around 170 bushels per acre [27]. Every bushel can produce around 2.8 gallons of ethanol [28]. The net result is that if we used all of the corn grown in the United States to make ethanol, we could produce around 41 billion gallons of ethanol annually.

While this may sound like an enormous number, the United States burned around 140 billion gallons of gasoline in 2015, more than three times this amount [29].

In addition, ethanol has a lower energy content per gallon than petroleum-derived gasoline does (roughly 50 % less, though mileage is typically around 75–80 % of what it would be with regular gasoline) [30]. Thus, it would require perhaps 180 or 190 billion gallons of ethanol to completely replace gasoline as a fuel for cars in the United States.

There is one more issue, and it is extremely pressing – a significant fraction of the corn grown gets eaten, either by animals or people. In fact, farming on this scale and then using all of the grain to produce ethanol would result in massive food shortages!

Biodiesel has similar problems. The United States used around 36 billion gallons of diesel for on-road transportation in 2014 [31]. This does not include fuel used by tractors, trains, and other off-road vehicles. In 2015, the United States. produced around nearly four billion bushels of soybeans, which could produce around six billion gallons of biodiesel [32]. This is sufficient to replace just 17 % of the on-road diesel used each year. And again, most of the soybean crop was used to produce food for animals and people.

Solid biofuels have different but still important limitations. Typically, the biomass being used as solid biofuels is inedible to people, and thus does not contribute to food shortage or higher food prices. For example, corn cobs, stalks, and other parts of the plant are not eaten but could be burned to produce energy.

The two main problems with using solid biomass in this way are fuel quality and transportation. The inedible parts of a plant are typically a low-quality fuel that does not produce much energy when burned. In addition, it is spread out over millions of acres of farmland, but would need to be collected and transported to a power plant in order to be utilized as a biofuel.

Gaseous biofuels suffer from the same transportation problem – the biomass needed to generate a flammable gas are not concentrated in one area. Additionally, producing a gas from biomass is a slow process and involves storing the biomass for long periods of time. Not all of the biomass will be converted to methane, leaving behind waste products. These must be disposed of, and the energy and cost for this disposal must be considered.

Perhaps my greatest challenge as a high school chemistry teacher is helping students who are required to take the course. The problems with biofuels can provide a connection point for these students. The disadvantages of biofuels are economic, cultural, and aesthetic in nature. By relating biofuel disadvantages to fields unrelated to chemistry, teachers can provide a feeling of inclusion to students whose interests may not be primarily scientific.

7.7 Potential for future improvements in biofuels

Existing biofuels cannot replace fossil fuels outright. So, why are people so interested? The answer lies in the prospect of making biofuels from other sources, specifically waste. Biofuels have already been made from turkey offal (the portions of turkey that are not eaten), sawdust, forestry waste, even used cooking oil. But, these sources are relatively small. There is a much larger source of waste that, if it could be used, would allow biofuels to take over a significant fraction of energy production: cellulose.

When ethanol is made from grain, an enormous amount of energy, water, fertilizer, pesticides, and other inputs have to be used to grow the crop (in the United States, this is almost always corn). Then, the crop is harvested, and only the corn kernels are used. The stalks, leaves, roots, and husks are all discarded. The primary solid constituent of these waste products is *cellulose*, a long-chain polymer formed from many sugar molecules.

Currently, the technology to produce ethanol from cellulose exists, but processing the cellulose into sugar requires a significant amount of energy and very specialized enzymes. These restrictions limit the use of cellulose as a feedstock for the production of ethanol. However, it is estimated that over a billion tons of cellulosic waste is available annually in the United States, sufficient to replace around one-third of transportation fuel in the United States. This does not include cardboard, paper, and packaging waste, which currently accounts for around 40 % of municipal waste. Using marginal land to grow crops rich in cellulose (such as grasses) could increase production still further, allowing ethanol to become the dominant transportation fuel in the United States.

Similar advancements are possible in the production of gaseous biofuels. Improved catalysts can increase the yield of combustible gases produced from solid waste products. Genetic engineering of microbes is also being investigated.

Many hurdles remain if biofuels are going to become a major source of future energy production. However, with continued research and investment, biofuels can be an increasingly important tool in the fight against climate change and a high quality, domestically produced fuel source that the United States can rely on for decades to come.

References

[1] James SR, Dennell RW, Gilbert AS, Lewis HT, Gowlett JAJ, Lynch TF, et al. Hominid use of fire in the lower and middle pleistocene: a review of the evidence [and comments and replies]. Curr Anthropol. 1989;30 1, 1–26.

[2] The Geography of Transport Systems. Jan 2017 Available at: https://people.hofstra.edu/geotrans/eng/ch8en/conc8en/worldenergyproduction.html. Accessed: 10 Jan 2017 XXXX.

[3] Use of NOAA ESRL Data. Jan 2017 Available at: ftp://aftp.cmdl.noaa.gov/products/trends/co2/co2_mm_mlo.txt. Accessed: 10 Jan 2017 XXXX.

[4] Earth Observatory. The effects of changing the carbon cycle. Jan 2017 Available at: http://earthobservatory.nasa.gov/Features/CarbonCycle/page5.php. Accessed: 10 Jan 2017 XXXX.

[5] Nasa: Warming seas and melting ice sheets. Jan 2017 Available at: http://climate.nasa.gov/news/2328/warming-seas-and-melting-ice-sheets/. Accessed: 10 Jan 2017 XXXX.

[6] EPA: Understanding the link between climate change and extreme weather. XXXX https://www.epa.gov/climate-change-science/understanding-link-between-climate-change-and-extreme-weather.

[7] NOAA: PMEL Carbon Program. XXXX ttp://www.pmel.noaa.gov/co2/story/What+is+Ocean+Acidification%3F.

[8] World Ocean Review: The battle for the coast. Jan 2017 Available at: http://worldoceanreview.com/en/wor-1/coasts/living-in-coastal-areas/. Accessed: 10 Jan 2017 XXXX.

[9] Saving Miami: A Stoichiometry Project. Jan 2017 Available at: https://docs.google.com/document/d/1G2jylu5VGyFPfR3uT-0FXGLOTyhRO_MptHfgXLQEA0M/edit#heading=h.uhvl2lidzeow. Accessed: 10 Jan 2017 XXXX.

[10] NREL: Lessons Learned from Existing Biomass Power Plants. Jan 2017 Available at: http://www.nrel.gov/docs/fy00osti/26946.pdf. Accessed: 10 Jan 2017 XXXX.

[11] Biofuels for transportation: Global potential and implications for sustainable agriculture and energy in the 21st century. Jan 2017 Available at: http://www.iinas.org/tl_files/iinas/downloads/bio/oeko/2006_Biofuels_for_Transportation-WWI.pdf. Accessed: 10 Jan 2017 XXXX.

[12] US Energy Information Administration: How much ethanol is in gasoline, and how does it affect fuel economy? Jan 2017 Available at: http://www.eia.gov/tools/faqs/faq.cfm?id=27&t=10. Accessed: 10 Jan 2017 XXXX.

[13] ChemBioEng Reviews. Biodiesel: Sustainable energy replacement to petroleum-based diesel fuel – a review. Jan 2017 Available at: http://onlinelibrary.wiley.com/doi/10.1002/cben.201400024/abstract. Accessed: 10 Jan 2017 XXXX.

[14] Progress in Energy and Combustion Science. Biodiesel and renewable diesel: A comparison. Jan 2017 Available at: htttps://naldc.nal.usda.gov/download/39385/. Accessed: 10 Jan 2017 XXXX.

[15] Biofuels.org.uk. Biofuels: The fuel of the future. Jan 2017 Available at: http://biofuel.org.uk/syngas.html. Accessed: 10 Jan 2017 XXXX.

[16] ABCs of Biofuels. Jan 2017 Available at: http://www1.eere.energy.gov/bioenergy/pdfs/Archive/abcs_biofuels.html. Accessed: 10 Jan 2017 XXXX.

[17] Energy in Agriculture and Society: Insights from the Sunshine Farm. Jan 2017 Available at: https://landinstitute.org/wp-content/uploads/2016/09/EnergySSF_tables1.pdf. Accessed: 10 Jan 2017 XXXX.

[18] USDA: Energy Life-Cycle Assessment of Soybean Biodiesel. Jan 2017 Available at: http://www.usda.gov/oce/reports/energy/ELCAofSoybeanBiodiesel91409.pdf. Accessed: 10 Jan 2017 XXXX.

[19] Evaluating feasibility and sustainability of bioethanol production: A case study comparison in China (Wheat) and Italy (Corn). Jan 2017 Available at: http://www.cep.ees.ufl.edu/emergy/documents/conferences/ERC05_2008/ERC05_2008_Chapter_26.pdf. Accessed: 10 Jan 2017 XXXX.

[20] USDA and US Department of Energy: An overview of biodiesel and petroleum diesel life cycles. Jan 2017 Available at: http://www.nrel.gov/docs/legosti/fy98/24772.pdf. Accessed: 10 Jan 2017 XXXX.

[21] Estimating the net energy balance of corn ethanol. Jan 2017 Available at: https://www.ers.usda.gov/webdocs/publications/aer721/32459_aer721.pdf. Accessed: 10 Jan 2017 XXXX.

[22] Schmer MR1, Vogel KP, Mitchell RB, Perrin RK. Net energy of cellulosic ethanol from switchgrass. Proc Natl Acad Sci U S A. 2008 Jan 15;105(2):464–469. Jan15. Epub 2008 Jan 7. Available at: https://www.ncbi.nlm.nih.gov/pubmed/18180449/Accessed: 10 Jan 2017 DOI:10.1073/pnas.0704767105.

[23] Australian Journal of Crop Science: Genetic resources of energy crops: Biological systems to combat climate change. Jan 2017 Available at: https://pubag.nal.usda.gov/pubag/downloadPDF.xhtml?id=49241&content=PDF. Accessed: 10 Jan 2017 XXXX.

[24] Bioenergy and Sustainability. Feedstock supply chains. Jan 2017 Available at: http://bioenfapesp.org/scopebioenergy/images/chapters/bioen-scope_chapter11.pdf. Accessed: 10 Jan 2017 XXXX.

[25] Biodiesel Magazine: Biodiesel energy balance surpasses 5.5-to-1. Jan 2017 Available at: http://www.biodieselmagazine.com/articles/7948/biodiesel-energy-balance-surpasses-5-5-to-1. Accessed: 10 Jan 2017 XXXX.

[26] USDA, Acerage: Corn Planted Acreage Up 7 Percent from 2015; Soybean Acreage Up 1 Percent; All Wheat Acreage Down 7 Percent; All Cotton Acreage Up 17 Percent. Jan 2017 Available at: http://www.usda.gov/nass/PUBS/TODAYRPT/acrg0616.pdf. Accessed: 10 Jan 2017.

[27] USDA Crop Production. Jan 2015. Summary, Jan 2016 Available at: http://www.usda.gov/nass/PUBS/TODAYRPT/cropan16.pdf. Accessed: 10 Jan 2017.
[28] Corn Ethanol Production. Jan 2017 Available at: http://articles.extension.org/pages/14044/corn-ethanol-production. Accessed: 10 Jan 2017 XXXX.
[29] US Energy Information Administration. How much gasoline does the United States consume? Jan 2017 Available at: https://www.eia.gov/tools/faqs/faq.cfm?id=23&t=10. Accessed: 10 Jan 2017 XXXX.
[30] US Department of Energy. Ethanol. Jan 2017 Available at: https://www.fueleconomy.gov/feg/ethanol.shtml. Accessed: 10 Jan 2017 XXXX.
[31] American Fuels: Alternative Fuels News and Commentary. Jan 2017 Available at: http://www.americanfuels.net/2014/04/us-on-highway-diesel-fuel-consumption.html. Accessed: 10 Jan 2017 XXXX.
[32] University of Nebraska – Lincoln: Institute of Agriculture and Natural Resources, Soybeans. Jan 2017 Available at: http://cropwatch.unl.edu/bioenergy/soybeans. Accessed: 10 Jan 2017 XXXX.

Steven Kosmas

8 Growing your green chemistry mindset

Abstract: The purpose of this article is not to delineate the steps to move across the continuum to being a greener chemist, but to analyse the cognitive processes involved in fostering a green chemistry growth mindset (GCGM) [Dweck C. (2006) Mindset: The New Psychology of Success. New York, NY: Ballatine]. The focus is on changing the mindset, which inevitably will lead to a more mindful approach to chemistry practices before the laboratory begins. A green chemistry fixed mindset (GCFM) is closed to making improvements, since the attitude is that the techniques and processes in the laboratory are already employing a green chemistry mindset [Dweck C. (2006) Mindset: The New Psychology of Success. New York, NY: Ballatine]. The problem with the GCFM is that it precludes the possibility of making improvements. However, the GCGM employs a continuous, intentional focus on the attitude towards green chemistry, with the ultimate goal being a change in chemistry practices that is greener. The focus of this article will be on the GCGM.

Keywords: green chemistry, safety, sustainability

8.1 Introduction

Becoming a greener chemist is a process which involves changing one's attitude as well as an open mindedness. It is no longer acceptable to say, "I've been using this chemical for years and the amount of waste I produce is insignificant" when there is a better alternative. Developing a green chemistry growth mindset (GCGM) takes work [1]. The process involves rethinking demonstrations and laboratories for greener alternatives. There has to be a change in attitude in order to develop the GCGM. Operating out of older paradigms, such as "I am too busy to deal with GCGM," will not lead to a significant improvement. At the same time, being overly concerned and mired down in detail may lead to the inability to execute and promote our wonderful field of study (yes, chemistry). This end of the continuum is the equivalent to saying, "I can't improve, since I am too busy reading five books on self-improvement concurrently." Green chemistry is a very healthy, balanced process that leads to a more sustainable future. We will look here at three principles of green chemistry and see how they can be integrated into the high school class or laboratory. While not formal case studies, these three examples provide several spots in a traditional curriculum where a green chemical principle can be inserted without any major change in focus for the course.

https://doi.org/10.1515/9783110445923-008

8.2 Principle #5 Safer Solvents & Auxiliaries

"Principle #5: Designing Safer Chemicals: Minimize toxicity directly by molecular design. Predict and evaluate aspects such as physical properties, toxicity, and environmental fate throughout the design process."

Let's take a look at one of the principles of green chemistry, #5, Safer Solvents & Auxiliaries. The dry cleaning industry used perchloroethylene (PCE) for a number of years. The use of this solvent peaked in 1980 [2]. Many people in the business knew that a better alternative was necessary, but the GCGM was not in vogue at that time. Having spent time in my father's dry cleaners, my gut level intuition led me to believe that breathing PCE commonly called "perc" was probably not the best thing for my health. As time went on, my gut level intuition was supported, since the Environmental Protection Agency stepped in and suggested in 1990 that emissions of PCE be limited from dry cleaning plants [2].

Another relatively unknown fact is that Freon 113 (1,2,2-trichlorotrifluoroethane) was used as a solvent in dry cleaning until the Montreal Protocol prohibited its use due to its impact on the ozone in the stratosphere. Freon 113 was also used as a refrigerant, and very few people thought about ozone depletion when they were recharging their air conditioners. The GCGM requires one to look for alternatives (e. g. better solvents) before litigation occurs or before coercion by the EPA. This mindset is constantly asking: what is the better alternative compared to, it is still legal? Had a GCGM been in place perhaps Freon 113 and PCE would have been replaced sooner [2].

Another solvent used in dry cleaning is liquid CO_2, which must be under pressure, since dry ice (solid CO_2) changes directly from a solid to a gas at normal atmospheric pressure. This begs the question: when was the first commercial liquid carbon dioxide plant built in the United States? The answer is in 1999 in Wilmington, North Carolina [5]. This in turn begs the next question: would the first commercial liquid carbon dioxide plant have been built earlier if the GCGM was more prevalent at that time?

The intent of this discussion is to intimate or more realistically assert that being a green chemist is not a simple process. As soon as a chemist starts thinking about how to green a process, the GCGM has been initiated. However, like all great endeavours, the process is not always easy and at times may challenge what might be called one's frustration tolerance level.

The following three examples provide students with the opportunity to segue into a discussion in a high school chemistry class.

1. Solvents. Water is the most common solvent that high school students will encounter. By providing real-life examples of other solvents and how they are used in industry, students are given the opportunity to compare and contrast which solvent is best when studying the solvent from a green chemistry perspective.

2. Phase changes and vapour pressure. The phase changes of CO_2 can be demonstrated by placing small pieces of dry ice into a plastic tube that can be closed as the dry ice sublimes into the gas phase, the students will witness the formation of liquid CO_2 as soon as the pressure is a little over 5 atm. This provides the students with the opportunity to see a liquid solvent other than water and a phase change that is less familiar (i.e. sublimation). The vapours associated with PCE and Freon 113 (e. g. vapour pressures) can also be part of a classroom discussion. Students will deepen their understanding of chemistry by researching and determining how green each material is.
3. Organic nomenclature and structure. Teaching nomenclature is a very important objective of the high school chemistry class. I personally think students should treat an introductory chemistry class as a foreign language class 10% of the time. Students should be given the opportunity to verbalize and discuss nomenclature. Freon 113 and "perc" serve as excellent examples of chemicals that have a systematic name and a common name. This discussion could conclude with the actions that these two chemicals have in the atmosphere, and their effect on the environment.

Though the discussion will be multifaceted as students discuss these three examples, in the end the goal is to direct the discussion back to green chemistry.

8.3 Principle #4 Designing safer chemicals

"Principle #4: Designing Safer Chemicals: Minimize toxicity directly by molecular design. Predict and evaluate aspects such as physical properties, toxicity, and environmental fate throughout the design process."

This sounds like a thought process that an industrial chemist would use when designing a greener process. Extend this concept to the high school setting where the teacher deliberately chooses a demonstration based on whether the process is green. Once again, the focus will be on growing the GCGM. In this section, the teacher has decided to do an equilibrium demonstration. If the teacher chooses to use cobalt as indicated by the reaction below, then two questions arise. The first question most often asked is how to dispose of cobalt compounds. This question falls under the domain of environmental chemistry. One has created something that may pose a threat to the environment, what should be done at this point in time? The GCGM forces a teacher to consider whether the cobalt complex should be used in the first place. Though one may find this equilibrium expression intriguing or exceptionally interesting, the green chemistry question still remains: is this the best choice as an equilibrium demonstration? The teacher's goal is to go through the green chemistry process, not directly answer the question for the students. If silver nitrate is added to the solution, then some silver chloride (AgCl) will precipitate out moving the equilibrium to the reactant side and the solution will be pink (assuming the starting solution is blue), as shown in the reaction. This is fascinating, since the silver chloride precipitate has formed and the equilibrium has shifted to the left. At this point, acetone can be added which will pull some of the water molecules towards the acetone which will lead to the solution turning blue in that region. This is a classic demonstration that has been used for decades. At this point in time, the question still remains: is this the best equilibrium expression to use to demonstrate the concept? The silver chloride precipitate can be filtered out, but this still leaves the green chemistry question unanswered. Is this equilibrium process green?

$$\left[Co(H_2O)_6\right]^{2+}(aq)(pink) + 4Cl^-(aq) \rightleftharpoons [CoCl_4]^{2-}(aq)(blue) + 6H_2O(l)$$

Though the chemistry involved in this equilibrium demonstration is extremely interesting, maybe it is best to use a video and limit the amount of waste produced on a yearly basis. This is a form of greening of the classroom that may be useful, but that should be noted to the students. Videos can enhance classroom learning and save time while greening the curriculum. Informational websites or videos should be incorporated into the curriculum when the teacher wants to green the curriculum, but not skip a great teaching opportunity [3, 4].

Let's continue by examining a second equilibrium demonstration. In this demonstration, copper sulphate pentahydrate is dissolved in water forming a pale blue solution. At this point in time, this demonstration is not spectacular at all unless students are shown the copper sulphate pentahydrate crystals which are universally accepted as being "very cool" crystals based on their appearance (my bias was added intentionally). If concentrated ammonia is added to the solution, then the dark blue copper ammonia complex forms. To simplify the chemistry, this equilibrium expression has the water molecules removed. I highly suggest that this demonstration is done in a flat-bottom Florence flask. When concentrated HCl is added, a cloud of NH_4Cl solid is produced above the aqueous phase. This precipitate baffles students, since the ammonium chloride precipitate forms while the equilibrium shifts to the left. The removal of the ammonia (NH_3) occurs when the precipitate forms shifting the equilibrium to the left and a pale blue-coloured solution exists in the flat bottom Florence flask. Add additional concentrated ammonia and the dark blue copper ammonia complex $Cu(NH_3)_4^{2+}(aq)$ will form again as shown in the reaction, below.

$$\underset{\text{pale blue}}{Cu^{2+}_{(aq)}} + 4NH_{3(aq)} \leftrightarrow \underset{\text{dark blue}}{Cu(NH_3)_4{}^{2+}_{(aq)}}$$

The question in this case is: is this copper ammonia complex greener than the cobalt complex? Remember that the teaching goal is not to answer the question, but to model the thinking process to empower the development of the GCGM. GCGM may lead some to discuss the fumes from this demonstration, since breathing even small amounts of concentrated ammonia is not an enjoyable experience. Some may argue that using copper is greener than using cobalt.

These two equilibria reactions again provide examples of how green chemistry can be incorporated into the classroom. Examples include:

1. Is one particular equilibrium reaction greener than another?
2. Does the presence of a precipitate have any correlation to the "greenness" of a reaction?

Both questions can become sources of a larger discussion that include how one ion or another exists in solution, and if the use of one material is inherently safer than another, lessening the toxicity of the reaction.

8.4 Principle #12 Safer Chemistry for Accident Prevention

"Principle 12: Safer Chemistry for Accident Prevention: Choose and develop chemical procedures that are safer and inherently minimize the risk of accidents. Know the possible risks and assess them beforehand."

Freon 113, as seen in Figure 8.1, as indicated earlier, had a negative impact on the ozone. Most cars used Freon 113 as a refrigerant until companies switched over to the new refrigerant. One can make the argument that the environmental mindset and the green mindset had very little to do with the switch. The change was motivated by economics. The new refrigerant was cheaper than Freon 113. If one had a GCGM back when people were changing which refrigerant that was used in their car, then an economic incentive might not have been needed.

It doesn't appear that many people are going to be concerned about what 1,1,2-trichloro-1,2,2-trifluoroethane (Freon 113) is doing to our stratosphere. Recalling from my personal observations at the time, many people interpreted the change in refrigerants as a nuisance and as unnecessary. Yet, this molecule is broken down to chlorine radicals which then break down ozone. Many students in a middle school

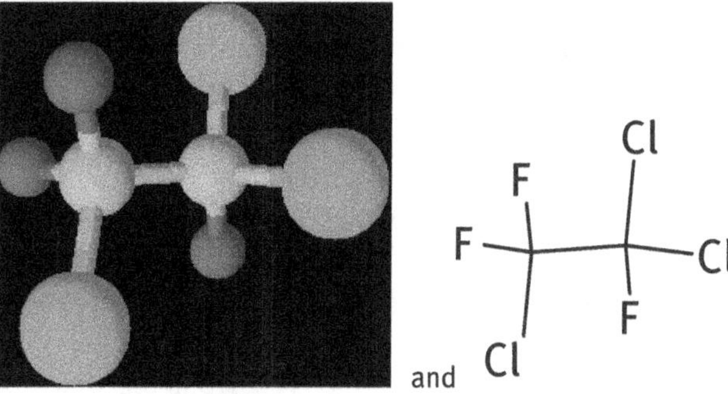

Figure 8.1: Images of Freon 113, 1,1,2-trichloro-1,2,2-trifluoroethane.

or high school setting may find this information interesting as part of a classroom discussion.

If GCGM was part of mainstream thinking, then more people may ask whether basic chemicals are safer for the environment or not and inherently minimize risks. For example, ethylene glycol is used in the radiator of cars to lower the freezing point of the solution in the radiator. Yet, how many people are asking whether or not this is the best green alternative? One can imagine that by shifting our thinking in the green direction, we can impact a large segment of the society, at least beginning with students in high school chemistry classes. If a larger percentage of people ask for greener alternatives, then according to supply and demand economics more alternatives will be made available. Currently, many people are using ethylene glycol as a mixture with water in their coolant system. This may seem extreme, but using 100% water may be the greenest alternative in areas where the temperature never goes below 0°C (32°F).

Additionally, using a GCGM, can an argument be made that propylene glycol is a better alternative? Ethylene glycol shows up in forensic videos as a poison, and if it is accidentally spilled on a driveway, cats and dogs may consume a small amount due to the sweet taste (as a disclaimer: I have never taste tested this poison, but it appears to be well known that ethylene glycol has a sweet taste). If every teacher discussed the advantages of using propylene glycol as a coolant in the radiator of a car, then at the very least there would be a discussion of the toxicity difference between the two chemicals which could model the thinking involved in fostering the development of a GCGM. If it can be assumed that some people will spill some chemical when they are adding antifreeze to their radiator, then some number of people may choose to use propylene glycol. Principle number 12: Safer Chemistry for Accident Prevention, will be supported if more people choose a safer coolant for their radiator. The GCGM will create more discussion and if more teachers are given training in this area, then each individual student will have heard about it from their teachers.

Once again, this topic becomes one that can easily be incorporated into the classroom. Students are usually quite familiar with automobiles, and some students are fascinated by NASCAR and drag racing, so by bringing examples of automotive chemistry into the lecture or the laboratory student interest can be piqued.

8.5 Conclusions

At this point in time, it is imperative that teachers (K-16) start discussing these concepts with their students. The first step is for the teacher to develop a GCGM. The principles that we have studied involve predicting and evaluating the process before implementation. This requires work, and it will not always be fun. Going back to the two equilibrium demonstrations, one hopes you are trying to weigh out the different possibilities between using “wet chemistry” versus video or which equilibrium demonstration should be chosen and why. If you have a rationale for what you are doing from a green chemistry perspective, then you have probably initiated the GCGM.

The GCGM is a process and one can become concerned that teachers will not be given the time or the opportunity to grow in this area, since most people cannot explain the difference between an environmental mindset and a green mindset. Delineating them may lead to an opening activity where students in class or an online forum debate the difference between an environmental mindset and a green mindset. From that starting point, adding examples to the class can be done without difficulty.

References

[1] Dweck C. Mindset: The new psychology of success. New York, NY: Ballatine. 2006. ISBN: 978-0-345-47232-8.

[2] https://drycleancoalition.org/chemicals/chemicalsusedindrycleaningoperations.pdf.

[3] Chem Demos, University of Oregon. Jan 2017. Available at: http://chemdemos.uoregon.edu/demos/Copper-Ammonia-Complex. 9 Jan 2017.

[4] Learn Chemistry: Enhancing Learning and Teaching. Jan 2017. Available at: http://www.rsc.org/learn-chemistry/resource/res00000001/the-equilibrium-between-two-coloured-cobalt-species?cmpid=CMP00005957. 9 Jan 2017.

[5] Wentz M, Beck DR, Monfalcone III V. Colorfastness of fabrics cleaned in liquid carbon dioxide Vol. 1. Research Triangle Park: American Association of Textile Chemists and Colorists, 2001(5).

Index

https://doi.org/10.1515/9783110445923-009

Also of interest

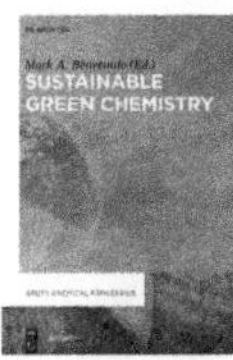

Volume 1
Sustainable Green Chemistry.
Benvenuto (Ed.), 2017
ISBN 978-3-11-044189-5, e-ISBN 978-3-11-043585-6
ISSN 2366-2115

GREEN – Alternative Energy Resources
Volume 1
Pyrolysis of Biomass.
Wang, Luo, 2016
ISBN 978-3-11-037457-5, e-ISBN 978-3-11-036966-3
ISSN 2509-7237

Sustainable Chemical Production Processes.
Marin, van Geem, 2018
ISBN 978-3-11-026975-8, e-ISBN 978-3-11-026992-5

Environmental Nanoscience.
Implication of Anthropogenic Nanomaterials
Obare (Ed.), 2018
ISBN 978-3-11-034234-5, e-ISBN 978-3-11-034235-2

Biorefineries.
An Introduction
Aresta, Dibenedetto, Dumeignil (Eds.), 2015
ISBN 978-3-11-033153-0, e-ISBN 978-3-11-033158-5

Sustainable Process Engineering.
Prospects and Opportunities
Koltuniewicz, 2014
ISBN 978-3-11-030875-4, e-ISBN 978-3-11-030876-1